LEIBNIZ: REPRESENTATION, CONTINUITY AND THE SPATIOTEMPORAL

Science and Philosophy

VOLUME 7

This series has been established as a forum for contemporary analysis of philosophical problems which arise in connection with the construction of theories in the physical and the biological sciences. Contributions will not place particular emphasis on any one school of philosophical thought. However, they will reflect the belief that the philosophy of science must be firmly rooted in an examination of actual scientific practice. Thus, the volumes in this series will include or depend significantly upon an analysis of the history of science, recent or past. The Editors welcome contributions from scientists as well as from philosophers and historians of science.

DIONYSIOS A. ANAPOLITANOS
University of Athens, Greece

LEIBNIZ: REPRESENTATION, CONTINUITY AND THE SPATIOTEMPORAL

Springer-Science+Business Media, B.V.

A C.I.P. Catalogue record for this book is available from the Library of Congress.

ISBN 978-90-481-5138-7 ISBN 978-94-015-9147-8 (eBook)
DOI 10.1007/978-94-015-9147-8

Printed on acid-free paper

**Dedicated to the memory of
Wilfrid Sellars**

TABLE OF CONTENTS

PREFACE ix
ABBREVIATIONS xii

CHAPTER I / REPRESENTATIONALISM 1
A. THE GENERAL SETTING 1
B. REPRESENTATIONS AND RELATIONS 12
C. INDIRECT REPRESENTATIONS 20
D. PERCEPTION 25
E. DISTINCT, CONFUSED, PETITE, AND UNCONSCIOUS PERCEPTIONS 31

CHAPTER II / CONTINUITY 50
A. THE PRINCIPLE OF CONTINUITY AND ITS RELATION TO OTHER LEIBNIZIAN PRINCIPLES 50
B. KINDS OF CONTINUITY AND APPARENT CASES OF DISCONTINUITY 63
C. DENSITY AND SEQUENTIAL OR CAUCHY COMPLETENESS 69
D. THE ROLE OF CONTINUITY IN THE FORMATION OF THE LEIBNIZIAN SYSTEM 78

CHAPTER III / SPACE, TIME, THE SPATIO-TEMPORAL, AND MONADIC REALITY 94
A. SPACE, THE SPATIALLY EXTENDED, AND ITS METAPHYSICAL CORRELATE 94
B. THE PHENOMENAL RELATION OF "SPATIALLY BETWEEN", ITS METAPHYSICAL BASIS, AND DENSITY AS A PROPERTY OF THE SPATIALLY EXTENDED 121
C. TIME, PHENOMENAL CHANGE, AND MONADIC CHANGE 134
D. FORMALIZATION OF THE THIRD MODEL AS IT REFERS TO DENSITY OF MONADIC AND PHENOMENAL CHANGE 153
E. ACTUAL INFINITY AND THE LEIBNIZIAN SOLUTION TO THE PROBLEM OF THE COMPOSITION OF THE CONTINUUM 158

BIBLIOGRAPHY 175
INDEX 183

PREFACE

The goal of this work is to articulate and defend a comprehensive new interpretation of Leibniz's later views concerning representation, continuity, and the spatio-temporal. Although the literature on Leibniz concerning these issues is quite rich, no serious attempt has been made to give a unified treatment of them. The interpretation proposed and defended is meant to fill the existing gap. Representation is a key notion for the correct interpretation of Leibniz. Strangely enough, despite its extensive use by Leibnizian scholars, an important subspecies of it, *indirect representation*, has been largely ignored. In light of Wilfrid Sellars' work, I came to realize that indirect representation is crucial for the understanding of Leibniz's metaphysics of the continuous and the spatio-temporal. Along with the rest of Leibniz's representational concepts (such as, e.g., those of distinct, confused, unconscious, and petite perceptions) indirect representation, appropriately used, yields an interesting interpretation of Leibniz's later views on continuity, space, and time. Additionally, it articulates and brings to the fore the appropriate interconnections between the Leibnizian realms of the real, the phenomenal, and the ideal.

The first chapter is devoted to discussion of various aspects of Leibniz's representational metaphysics. Among other things, the notion of indirect representation is discussed in detail in its own right and the representational framework of the continuous and the spatio-temporal is established. The view is adopted that Leibniz has a tri-level system, including the level of the real (or metaphysical), of the phenomenal, and of the ideal. Special attention is paid to the problem of what sort of correspondence exists between the metaphysical and the phenomenal level in general.

In Section I.A the distinction, common to the Cartesian tradition, between formal and objective reality is briefly discussed. More specifically, it is argued that Leibniz tacitly employed this distinction, but modified it so that there was no longer a second distinction between primary and secondary qualities. In Section I.B, I adopt, discuss, and expand the Sellarsian interpretation of Leibniz's reduction of phenomenal relations to representational facts of the world of monads as they represent one another. In Section I.C the notion of indirect representation, which, I argue, has to be taken seriously if one wants to give a coherent interpretation of Leibniz's system, is defined and examined. Leibniz holds that what really is in the world is monads, i.e., substantial undivided units in their different representational states. So monads, by representing the world, represent all the other monads as they represent. In Section I.D the view is presented and defended that monads (and therefore "vulgar" individuals like human beings, animals, plants, chairs, etc.) are continuously graded according to the distinctness or confusedness of their representational structures. In Section I.E, I discuss in more detail distinct and confused perceptions, and present a novel account of petite and unconscious ones which ties up with my treatment of the continuity of what appears as spatially extended.

The second chapter deals with various aspects of Leibniz's notion of continuity. In Section II.A the most important architectonic principle of his system, the principle of continuity, is discussed. I consider two basic forms of it: (a) nature never makes leaps, and (b) the correlations between cases in nature-which can be considered as values of the same variable x-and their consequences are described by continuous functions. I prove that these forms are not equivalent, and propose a way out of this problem by an appropriate modification of (b), which, I argue, is permissible in the Leibnizian framework. Appropriate reconstructions of Leibniz's arguments for the derivation of this principle from the principle of sufficient reason and the principle of the best are also given. Additionally, it is argued that the principle of continuity is independent of the principle of the identity of indiscernibles and is not equivalent to the principle of plenitude. In Section II.B basic kinds of continuity and apparent cases of discontinuity are discussed as they appear in the framework of Leibniz's system. By using the distinct/confused dichotomy together with the Leibnizian doctrine concerning the finitude of what can be distinctly represented in the framework of one and the same representation, I establish that these apparent cases of discontinuity do not constitute a violation of the principle of continuity. In Section II.C the position is developed that not only density, but also traces of what we would today call sequential or Cauchy completeness can be found in Leibniz's writings. Finally, Section II.D contains an account of one of the basic frictions which, I think, led Leibniz to adopt his non-spatial, monadic metaphysics.

Chapter III discusses space, time, the phenomenally spatio-temporal, and monadic reality and change. In Section III.A an analysis of Leibniz's tri-level metaphysics as it refers to space (ideal), the spatially extended (phenomenal), and to monadic momentary representational reality (metaphysical) is offered. The property of *uninterruptedness* is critically assessed, especially as it refers to the continuity of space, considered as an ideal entity (and for that matter as it refers to any Leibnizian ideal continuum.) I argue that it is uninterruptedness that led Leibniz to consider the ideal ones as the only genuine continua and consequently to insist that he had solved the problem of the composition of the continuum. Section III.B is devoted to the construction of a model describing the interconnection of the real with the phenomenal in the case of the phenomenal relation "spatially between." In the course of the construction the notion of indirect representation is used. In addition, a formalization is proposed which leads to an exact account of spatial density in terms of its metaphysical correlate, representational density. In Section III.C I discuss time, phenomenal change, and monadic change. I present a model for phenomenal simultaneity as based upon metaphysical simultaneity. I argue that one is forced to admit monadic simultaneity as a feature of monadic change, if one wishes to give a coherent interpretation of Leibniz's representational metaphysics of change. This leads one to adopt the view that monadic change relations are in one sense real (connecting together metaphysical past and future) in a way in which metaphysical spatial relations are not. In the same Section three models of monadic change are considered under the names the *Whiteheadian Model*, the *Discrete* and *Discontinuous Model*, and the *Discrete* and *Continuous Model*. Arguments for and against them are presented, and the third is adopted. In Section III.D the Discrete and Continuous Model is used for a formalization of the interconnection of monadic and phenomenal change as it refers to temporal density. In Section III.E Leibniz's solution to the problem of the composition of the continuum is critically assessed and discussed. It is argued that Leibniz's solution was too easy a way out of the *labyrinthus continui* and, in a sense, too heavily dependent upon his idea of the uninterruptedness of ideal continua as the basic property of genuine continuity. It is concluded, finally, that without substantial modifications, his metaphysics could have allowed him to adopt a solution much closer to our modern conception of the problem.

I was very fortunate to have had the opportunity to discuss many of the ideas in this book with the late Wilfrid Sellars. I thank George Gale who read through a draft of the book and provided me with extremely helpful comments. I also have had many conversations with Nicholas Rescher, Hidé Ishiguro, James Edward McGuire, Joseph Camp and John Haugeland. I thank them all. Finally my thanks go to Nancy Nerssesian for her continuous encouragement while writing the book.

The responsibility for all translations of the quoted passages is mine, though I have made use of existing translations along the way, sometimes modifying them slightly. The following abbreviations are used throughout the text and refer to the corresponding items in the bibliography.

ABBREVIATIONS

A	[2]	L	[94]
AG	[7]	La	[92]
BC	[19]	M	[102]
Cout. OF	[24]	NE	[124]
F	[45]	R	[128]
G	[59]	Th	[43]
GM	[58]	W	[152]
J	[75]		

CHAPTER I

REPRESENTATIONALISM

A. THE GENERAL SETTING

It is more or less generally accepted that one of the basic themes of the philosophy of the 17th century was *representationalism*, that is, that our epistemological participation in the world is by way of our *representing* it. We can say confidently that a common implicit assumption of the philosophical systems of the time was the existence of a domain of *representables*, posterior to and dependent upon the necessary existence of God. We can even contrast it with the Platonic realm of Forms (or Ideas), of which it is in a certain sense a descendant. It is of course quite correct to note that the ontological status of the do*main* of Platonic Forms was different, at least insofar as they were not dependent on God. For the Christian thinkers of the 17th century, on the other hand, the realm of representables had to be ontologically dependent upon God. Whether the sort of dependence Descartes had in mind was different from the one that either Spinoza or Leibniz or anyone else at the time adhered to is an issue that does not concern us here. What is important is the fact that, putting aside differences of detail, the domain of representables was thought of as ontologically posterior to the existence of the supreme Being.

Representables exist, in their primary mode of being, as actually thought of by God. One way to understand this is to take the essence of them to be "being conceived by God." It is quite clear that God's acts of thought are dependent upon God, but are not taking place in time. Representables, so to speak, are not created one after another in a series of divine contemplations. They are everlasting and God's thinking of them is, in a radically non-temporal sense, what constitutes their essence.

Representables are in another, very definitely temporal sense two-faced. They can be realized in the world either formally or objectively or both. A representable, triangularity for example, can be formally realized in the world as a material triangle and objectively in the mind of Jones as a "being perceived" or, more generally, as a "being thought of" triangle. Other representables such as, e.g., a unicorn can also be objectively realized. A unicorn could possibly be formally realized also, but such a creature has never been observed in the real world, which leads us to think that its objective realization in us is a part of our imaginative mythological heritage. By using the expression "a unicorn could be formally realized" we mean that there is no logical inconsistency involved in thinking about such a possibility.

The situation is quite different when we think, for instance, of a diangle, a two-angled enclosed plane figure in a Euclidean space. It is true that we cannot form a picture of a diangle, but we can still think of it using some kind of analogy, that is, thinking of it as something analogous to a triangle, quadrangle, etc., but differentiated from them by the fact that it only possesses two angles. Such a representable can never be formally realized in a Euclidean world, because such a realization would be incompatible with the laws of Euclidean geometry. The objection that we know by now that this world is not Euclidean cannot be sustained, because, whatever the nature of our world, we can always find an example of a representable which cannot be formally realized because of its incompatibility with the geometrical or other laws of this world.

Finally, we can think about the possibility of representables formally realized of which we cannot have a mental counterpart, just because their formal realization took, is taking, or will take place in time or space so far away from any possible representor that this makes such a mental counterpart difficult to form. But even such a situation as this does not exclude the possibility of an objective representation of what formally exists and has not yet been observed, in future imaginative thought. Additionally, we could adopt the Leibnizian alternative, according to which we represent whatever *was*, *is*, or *will be* formally existing in the world, no matter whether or not we are conscious of having such a representation.

Cartesian representationalism was based upon the two-faced realization of representables. Triangularity, as we have said, could be realized in the spatio-temporal world by a material triangle and in the mind of Jones as a "being perceived" or, more generally, as a "being thought of" triangle. The Cartesian mechanism of perception involved a causal chain in which a corresponding modification of the pineal gland was the last stage before the idea of a triangle could be formed in the mind of Jones. The correspondence between the material triangle and the particular state of Jones' pineal gland could be physically explained through an appeal to a causal process, because, after all, both a physical triangle and Jones' pineal gland belonged to the same part of the world, namely, to that part about which physics provides us with convincing explanatory schemata concerning causal interconnections of its members. But the *right* correspondence between the state of Jones' pineal gland and Jones' perception of a triangle could only be explained through metaphysical reasoning. For the Christian metaphysicians of the 17th century, it was not enough to say that such a correspondence just happens to be the right one; in other words, it was not enough to say that a neuro-physiologically complete description of Jones' brain state was the most one could ask for as an explanation of the occurent mental state of Jones.

The epistemological gap between formal and objective reality was therefore bridged through an appeal to a rich ontology. It is God who has created the world so that formal and objective reality are two faces of the same coin, i.e., He guaranteed that they are *different* realizations of the *same* representable, the *essence* of which is "being conceived by God." Representables, in other words, are the ultimate causes for the *right* correspondence between formal and objective reality. For the 17th century thinkers, no matter what the mechanism of perceiving is, the guarantee for the final stage of a correct representation, provided that mistakes of a hallucinatory nature do not obtain, is the realm of representables as "being thought of" by God and the fact that God is not a deceiver. That objective and formal reality do agree and correspond has to be explained via the fact that they are realizations of the *same* representables. Representables as "being thought of" by a non-deceiving, benevolent God was the ground of the 17th century representational theory of truth. In other words, the contention was that humans can rest assured that what they perceive in normal circumstances is correct, because it is one of the two sides of the same coin, the other side being the formal realization of the same representable. This ontological causality[1] is the key to the solution of the mind-body problem for the Christian thinkers of the 17th century.

The Cartesian contrast between objective and formal reality was the main theme for the representational theories of truth which dominated the scene in 17th century philosophy. Objective and formal reality can be thought of, on the one hand, as by themselves forming a representational pair and, on the other, as being the end-points of a two-fold representation or realization of the representables as they exist, in a non-temporal sense, in God's mind. The question is, of course, what is representation and how we can specify the structural properties or qualities which can be thought of as characterizing both objective and formal reality. To examine Descartes' distinction between two sorts of perceptions, those "which have the soul as a cause and those which have the body [as a cause]" ([65] volume I, p. 340) is beyond the scope of this work. Additionally, we will ignore perceptions of a hallucinatory nature, and thoughts of angels, unicorns, and the like. The paradigm we will use in our discussion is that of correctly perceiving a triangular object as existing in the spatio-temporal world which is present out there independently of us.

---

[1] Or rather be-causality, if we use Sellars' terminology.

According to Descartes, the term *objective reality* of an idea[2] can be understood as signifying the existence of a certain mental counterpart of that which is represented. By using the traditional metaphor of containment he thinks of such a counterpart as the *objective content* of the idea. To it corresponds *formal reality*, that is, the object of the representation as existing in the spatio-temporal world. When we correctly perceive a triangular object as existing in the spatio-temporal world, we come to form an idea such that the entity which is its content has structural characteristics in a sense isomorphic to those of the triangular object. Such characteristics are of a geometric or mechanical nature. It seems that for Descartes the representational correspondence between objective and formal reality was analogous to the representational correspondence between an object as it exists (or as it seems to us to exist) in the spatio-temporal world and its depiction in the framework of a realistic painting. A correct perception of a triangular object contains, as it were, an entity which corresponds in its mental geometrical and mechanical features to the triangular object that is represented in the perception. As we said, this correspondence is in a way similar to that which obtains between the depiction of an object in a realistic painting and the object so depicted. There is an important epistemological difference between them, though. We can judge the correctness of the representation of the object in the painting. We could even have helped to have created it, if we had produced the painting ourselves. We can compare that with our inability to know whether there is a correct correspondence between formal and objective reality, even if we assume that the existence of the former is beyond doubt and that some kind of correspondence between them obtains. On the other hand, we can see the generic value that such an example has for the formation of a representational metaphysics of the Cartesian sort.

The truthfulness or the correctness of the correspondence between objective and formal reality is guaranteed by God. Both the triangular object as it formally exists and as it is represented are realizations of the same representable existing in God's mind. God is the source of truth. As a result, what is for him knowledge is for us metaphysical speculation, because empirical access to such a source is not available to us. God has created and can see both, us as representors and the world as we represent it. He has, as it were, immediate knowledge of the correctness or truth of such a representation. We were only given access to our representations and therefore we can only metaphysically postulate the existence of a spatio-temporal world

2 [65], volume II, pp. 52-53.

independent of us. We can, using the analogy of the painting and the object that is depicted in it, speak of a similar correspondence between what formally is the case and what we objectively take it to be. We are helped in that by criteria which we form out of the experienced internal coherence that separates what we come to think of as real phenomena from imaginary ones. Such an experience is valuable in the sense that without it we, quite probably, would be unable to build any metaphysics representational or not.

Leibniz formed his system taking for granted the Cartesian distinction between formal and objective reality. Of course, if we go to the Leibnizian corpus (especially to his mature writings) we do not find explicit references to the Cartesian representationalism as based upon the two-faced realization of representables. Yet there is a sense in which the basic Cartesian representational characteristics are present in the Leibnizian system. On the other hand, there is an important difference between Descartes' representational metaphysics and Leibniz's. It concerns the quite celebrated distinction between *primary* and *secondary* qualities. The Cartesian view was that the primary qualities were properties common[3] to the things as they really are and to the objectively existing entities in our perceptions (or better in our ideas) which correspond to them. He considered as his paradigmatic case shape and therefore spatial extension. That is, he thought that when one correctly perceives a spatial object as triangular, it is not only the case that there is something out there which roughly corresponds to one's mental image of it, but also that this something is extended and has a definite shape which is triangular in the way one perceives it.

Concerning secondary qualities the traditionally paradigmatic case was that of color. According to Descartes, qualities like color were in a fundamental way not *really* in the *object*. The idea lurking behind such a position was that only *mechanical* (which meant for him basically *geometrical*) qualities could be primary. In an enlightened century the idea that the only properties things *really* have are the properties necessary for scientific explanations came as no surprise. In addition, it was a firm 17th century belief that scientific explanations had to be mechanical. So shape was considered to be the paradigm of a primary quality. Spatial properties such as being triangular were, in other words, highly qualified for being primary. They were considered as literally *inherent in the object*. The secondary qualities were thought to be propensities that objects have to cause certain sensations in the observer, for

---

[3] This is not to say that the objectively existing entities in our perceptions share the property of being spatio-temporal with the formally existing things. On the other hand, spatio-temporal characteristics of the latter are correctly depicted by picture-like characteristics of the former.

instance, visual sensations of red, blue, etc. These propensities were not of any mysterious nature. They could be explained mechanically as being due to the object's texture, which reflects or refracts (or both) light in certain ways. The object's texture should be, at least in principle, completely describable in mechanical terms. Mechanics, in other words, was conceived as the final reductionist explanatory frontier for secondary properties.

Leibniz differs sharply from Descartes in his treatment of the primary/secondary dichotomy. For him both shape and color are phenomenal. There is neither color nor extension in formal reality. Formal reality consists in representors being in changing representational states. In formal reality there are configurations of representors in specific representational states which correspond in a coherent way to the perceived properties of being shaped or colored. But color or shape as such are not part of that reality. Color can still be explained mechanically, but the existence of such a scientific explanation is due to the nomological uniformities that connect phenomenal entities together. These nomological uniformities are based upon definite uniformities among what really exists, and what really exists are representors in particular representational states. The real world is neither colored nor spatial. The phenomenal world is both. There is a structural correspondence between them that makes scientific explanations parasitic upon metaphysical ones, such that the former can be clearly seen to belong entirely to the objective reality of the well-founded phenomena, the well-foundedness of which is a basic Leibnizian metaphysical doctrine.

There are two distinct[4] levels constitutive of the Leibnizian representational metaphysics. In the first we find a world exclusively populated by simple substances (monadic representors, each one representing this same world from its point of view.) They do not interact and are self-sufficient. They contain in their individual complete concepts once and for all everything that has happened, is happening, and will happen to them, and they are the centers of every activity, both metaphysical and phenomenal. In the second we find the world of well-founded phenomena whose nature is representational in the sense that they are the contents of the representational states of the monadic representors as they do perpectively represent what is going on in the real world of monadic representors.

[4] In fact, as we shall later see, there are three levels, those of the ideal, the phenomenal, and the real. Here we will be primarily concerned with the second and the third. In the first belong all the ideal entities, such as space, time, number, etc., which lack a unitary and substantial referent, and play the role of the abstract fabric our theoretical framework for describing the general features the world is made of.

In a letter to De Volder, dated June 30, 1704[5], Leibniz states that all there is in the world is simple substances moving from past to future perceptual states via an internal principle of spontaneous action. The disposition for such a spontaneous movement is called *appetition*. Appetition, in other words, is an unconscious active principle responsible for the perpetual change of the representational states of each particular monad. According to Rescher:

> ... each individual substance is subject to a perpetual continuous change of state, the only activity of which it is capable. Leibniz chooses to call it appetition, but connotations of this term toward some sort of active, conscious seeking or striving must be avoided[6].

The real cause of this monadic (metaphysical) activity is the internal, unconscious disposition to move toward the future, which is implanted in monads by God. The representations of all of these representors form a coherent and harmonious structure, pre-established by God. A representor does not represent in a vacuum, although such an alternative is not logically impossible. What makes it impossible is simply the fact that God is not a deceiver. The first level of the Leibnizian representational system, in other words, concerns the whole actual world as it really is, namely, an assemblage of monads in particular representational states. The second level is in a certain sense parasitic upon the first one. It concerns the world not as it is, but as it is faithfully represented from a particular perspective by each particular monad. What each such monadic representor represents is other representors as representing other representors as representing ... and so on *ad infinitum*. This kind of combination of direct and indirect representations together with the active principle of moving from one representational state to the next is all that is metaphysically real in a monad.

In the Leibnizian system we can find all the basic characteristics of the representational schema that was common to the philosophical systems of the 17th century. As we have already said, explicit references to Cartesian representationalism as based upon the two-faced exemplification of representables are not to be found in the Leibnizian writings. He does not talk explicitly about the distinction between formal and objective reality. Yet such a distinction not only exists in his system, but also becomes immediately apparent to an observant reader. The following two quotations from his mature writings bear witness to this assertion:

---

5 G, II, 270.

6 [125]. p. 86.

> *Simple substance* is that which has no parts. *Compound substance* is a collection of simple substances or *monads*...
> 2. *Monads*, having no parts, can neither be formed nor unmade. They can neither begin or end naturally, and therefore last as long as the universe, which will change but will not be destroyed. They cannot have shapes, for then they would have parts. It follows that one monad by itself and at a single moment cannot be distinguished from another except by its internal qualities and actions, and these can only be its *perceptions*-that is to say the representations of the compounds, or of that which is without, in the simple-and its *appetitions*-that is to say, its tendencies from one perception to another-which are the principles of change. For the simplicity of a substance does not prevent the plurality of modifications which must necessarily be found together in the same simple substance; and these modifications must consist of the variety of relations of correspondence which the substance has with things outside. (L, 636; G, VI, 598.)

In this first quotation (from *The Principles of Nature and of Grace Based on Reason)* one can recognize as *formal reality* the reality of monads together with their *internal qualities and actions* that is together with their perceptions (representations) and appetitions. In the following second quotation (from *The Monadology)* one can observe that what corresponds to the Cartesian objective reality in the Leibnizian metaphysics is the contents of each monad's representations which are perspective viewings of one *single universe*, i.e. of one and the same world of phenomena:

> 57. Just as the same city viewed from different sides appears to be different and to be, as it were, multiplied in perspective, so the infinite multitude of simple substances, which seem to be so many different universes, are nevertheless only the perspectives of a single universe according to the different points of view of each monad. (L, 648; G, VI, 616.)

As the first of the above two quoted passages makes clear, it is a firm doctrine of the Leibnizian metaphysics that what really exist in an absolute sense are monads in different and perpetually changing states. But what sort of things are the monadic states? Following the first quoted passage, we can rest assured that what makes a monad different from any other is its particular set of perceptions, which, needless to say, are representations from within of what is going on without. Additionally, we are told that all there is in a monad are its perceptions and its appetition. Monadic states are therefore representational states. So the real world consists of a discrete multitude of monads representing. Our next question is what it is that monads represent. They represent the world from their points of view. What constitutes the absolute reality of the world are indivisible substances (monads) in their different representational states. Each monad therefore represents the world by way of representing all of the other monads (distinctly or confusedly, directly or indirectly) as representing all the other monads (distinctly or confusedly, directly or indirectly) as representing ... and so on *ad infinitum*. In order to make sense of such a complicated notion of representation we will introduce later on (more specifically in Section I.C.) the notion of indirect representation as a necessary one for interpreting Leibniz important subspecies of the former.

To use the Cartesian terminology, formal reality is the world as it is, *namely*, inhabited by, or rather consisting of, representors in perpetually changing representational states. On the other hand, objective reality is the reality of the point-of-viewish representations of each particular representor. That these point-of-viewish representations agree with one another in normal circumstances is not to be explained by the existence of a world which is tied up with its representations through a kind of occasionalism or *influxus physicus*. For Leibniz, such a harmonious agreement had to be pre-established by God, the benevolent creator who had freely chosen to bring into existence this best of all possible worlds.

The doctrine that formal and objective reality are the two sides of the same coin, i.e., that they are realizations of representables existing in the mind of God, is also a Leibnizian doctrine. The following quotation from the *New Essays on Human Understanding* is quite revealing:

> God has ideas (of substances) before creating the objects of the ideas, and there is nothing to prevent him from passing such ideas on to intelligent creatures: there is not even any exact demonstration proving that the objects of our senses, and of the simple ideas which our senses present to us are outside of us. (NE, 296; G, V, 275.)

The passage which follows *(The Leibniz-Clarke Correspondence,* Leibniz's "Fifth Paper") is of additional value, since one can easily see that the ontological priority of representables in God's mind is a central doctrine of Leibniz's metaphysics.

> ... God himself cannot perceive things by the same means whereby he makes other beings perceive them. He perceives them, because he is able to produce that means. And other beings would not be caused to perceive them, if he himself did not produce them all harmonious, and had not therefore in himself a representation of them; not as if that representation came from the things, but because the things proceed from him and because he is efficient and exemplary cause of them. He perceives them, because they proceed from him; if one may be allowed to say, that he *perceives* them, which ought not to be said unless we divest that word of its imperfections; for else it seems to signify, that things act upon him. They exist and are known to him because he

> understands and wills them; and because what he wills, is the same as what exists. Which appears so much the more, because he makes them to be perceived by one another and makes them perceive one another in consequence of the natures which he has given them once for all, and which he keeps up only, according to the laws of every one of them severally; which though different from one another, yet terminate in an exact correspondence of the results of the whole (A, 84; G, VII, 411).

With the help of these passages one can distinguish at least four important themes which appear again and again in Leibniz's writings.

1. Representables are ontologically prior to their realizations. According to Leibniz.. "God has ideas... before creating the objects of these ideas." Furthermore, he communicates these ideas to intelligent creatures. Communicating ideas simply means that he makes it possible for appropriate realizations of his ideas to appear as objective counterparts (in the minds of his creatures of the formal reality of the structured world of objects he has also created.)

2. The existence of representables in God's mind is independent of the possible realizations of them. God knows and understands them as such in a timeless and unified way. Representables form the realm of *possibilia*, in the sense that their realizations depend upon God's free will. Following Rescher, we can say that according to Leibniz:

> Anterior to the existence of our world there was recorded in the divine mind entire infinities of notions of possible individual substances, whose only being at this point is that *sub ratione possibilitas* in God's mind[7].

In the Leibnizian passages we quoted above it is clear that Leibniz does not draw the distinction between representables realized in this world and representables not realized in it. Furthermore, he does not talk about compossibility as the necessary prerequisite for any world to be possible. There are other places where he does so[8]. It is nevertheless obvious that he draws a distinction between God's understanding and God's will. God understands and knows all possible representables which by way of being compossible or not are grouped together into different worlds. He wills to create this world which is the best possible. So "what he wills, is the same as what exists." The act of creation, on the other hand, is not an act of creating representables. It is an act of choosing the best possible lawful assemblage of compossible representables to be realized both formally and objectively.

---

7 [125], pp. 16-17.

8 See, e.g., G, III, 573.

3."...there is not even any exact demonstration proving that the objects of our senses, and of the simple ideas which our senses present to us are outside of us." This statement concerning our inability to prove by way of reasoning the existence of an external (formal) reality presents us with a classical and thorny philosophical problem. How, after all, can we be sure about the existence of such a reality of objects corresponding to our perceptual or mental images, by way of reasoning? Reasoning involves making comparisons and comparisons can be made among things of the same nature. How can we ever-to use Berkeley's terminology and argument-compare ideas with objects, if the only things we can have in our minds are ideas?

4. The way out of this problem can be achieved only through metaphysical reasoning. The answer to it has to come through the basic doctrines of a philosophical system. For Leibniz: (a) The existence of an external world of monads representing is guaranteed by the fact that God could not be a deceiver creating only us and making us represent in a vacuum of other substances. What we represent within has a solid basis, a solid ground without. (b) Additionally, since what we represent is not responsible for our representation of it because every monad is windowless, self-sufficient, and does not interact with others, God made a two-fold provision concerning the truthfulness of representations. First, we do *correctly* represent within things existing outside of us because both the thing represented and the thing as represented are the two faces of the same representable, and, second, the representations of all the representors, in normal circumstances, are coherently tied together by pre-established harmony.

The following three passages, the first two from a letter of Leibniz to Arnauld, dated July 4/14, 1986 and the third from the *Discourse on Metaphysics* provide additional textual support for the interpretation that we just developed. In the first, Leibniz says that the full concepts of the individuals are, in a primary way, "thought of" by God. In the second, he speaks of the formal realization of those that God decided to create. He accordingly distinguishes "essences" as they "exist in the divine understanding" from their formal realization as involving God's free will. In the third he talks about God as constituting the guarantee that our perceptions are "real," i.e., faithful representations of the world of monads as they represent:

> As for me, I had believed that the full and comprehensive concepts are represented in the divine understanding as they are in themselves. (M, 54; G, II, 48-49.)

> ... the concepts of individual substances, which are complete and capable of wholly distinguishing their subjects, and which consequently embrace contingent truths or truths of fact, and the individual circumstances of the time, the place, etc., must also embrace in their concept taken as possible, the free decrees of God, also viewed as possible, because these free decrees are the principal sources of existences or facts; whereas essences exist in the divine understanding prior to the consideration of the will. (M, 55; G, II, 49.)

> ... only God constitutes the link or communication between the substances, and it is through him that the phenomena of the one meet with and agree with those of the others and that consequently there is reality in our perceptions. (L,324; G, IV, 458.)

B. REPRESENTATIONS AND RELATIONS

For Leibniz in his mature period, what really exist are substantial indivisible units (monads) representing the world from their particular points of view. Formal reality constitutes a structural whole, the building blocks of which are partless simples. In a quite interesting way (which will become apparent as we go on), what makes such a whole structured is not the existence of relations over and above the simples they connect together. Instead the ultimate undivided units contain all their predicates in such a way that relations are phenomenal and reducible to predicative constituent parts, namely properties. In a letter to Des Bosses, dated April 21, 1714, Leibniz describes his position as follows:

> I do not believe that you will admit an accident that is in two subjects at the same time. My judgment about relations is that paternity in David is one thing, sonship in Solomon another, but that the relation common to both is a merely mental thing whose basis is the modifications of the individuals. (L, 609; G, II, 486.)

Let us consider the relational statement:

Solomon is the son of David.

Such a statement can be analyzed into two predicative ones:

- Solomon (is the son of David.)

and:

David (is the father of Solomon.)

In the first case "Solomon" is the subject and "is the son of David" the predicate, and in the second case "David" is the subject and "is the father of Solomon" the predicate. We can then say that, according to Leibniz, Solomon is *phenomenally related* to David as son to father by way of containing the predicate "is the son of David", and also that David is similarly related as father to son to Solomon by way of containing the predicate "is the father of Solomon." There is not a real relation over and above David and Solomon with them as its endpoints; both subjects contain appropriate predicates which correspond and cohere with each other in such a way as to give rise to a phenomenal relation that ties Solomon and David together in a specific manner described by the initial relational statement:

Solomon is the son of David.

Adopting Sellars' general discussion of the above situation, we can analyze the relational statement

S_1 is R to S_2.

where S_1 and S_2 signify particular monads, as follows: Leibniz holds that for this proposition to be true there must be an R-to-S_2 inherent in S_1. The problem we are facing is how we can make sense of such an inherence, given that all that exists is representors and their representational states[9]. According to Sellars ([136]. p. 4):

> R-to-S_2 inherent in S_1 is interpreted as a representing of S_2 inherent in S_1, and Leibniz therefore interprets the sense in which S_2 is a 'part' of the R-to-S_2 inherent in S_1 as a matter of its being that which has objective or representative being in the representing which is R-to-S_2. According to this analysis, the truth of statements of the form S_1 is R_i to S_2 where R_i is prima facie a real relation, rests on facts of the form S_1 represents (in specific manner M_i) S_2, where, needless to say, the manner of representation M_i which corresponds to R_i and makes this relational fact a phenomenon bene fundatum is not what common sense has in mind when it uses the term 'R_i'.

The basic ingredients of the Sellarsian interpretation of Leibniz's contention that relations are reducible to predicates inherent in their subject can be summarized as follows:

---

[9] According to Leibniz "only indivisible substances and their different states are absolutely real" (M, 153; G, II, 119.)

(a) R-to-S_2 is inherent in S_1 in the sense that it is a representing by S_1 of some particular aspect of the world of monads as they represent. That aspect is determined only by *what there* is, namely, by the world of monads as they represent.

(b) S_2 is inherent in S_1 as a representing which is a "part" of the representing R-to-S_2. So, using Cartesian terminology, S_2 is inherent in S_1 as an *objective* or *representative* being in the middle of a more complex representing which is the R-to-S_2. It corresponds to the *formal* reality of the monad S_2 and belongs to the more complex representing R-to-S_2, which corresponds in its turn, by way of being their representations, to particular facts of the world of monads, namely, to particular complexes of concurrent representational states of the inhabitants of this same world.

(c) Relational truths rest on facts of a representational nature. Such facts are predicative and not relational.

(d) To the *prima facie* real relation R_i corresponds a specific *manner* of representation M_i which is defined by structural characteristics of a representational nature. We will see more of those characteristics when we examine the notions of direct and indirect representations, as well as those of confused, petite, and unconscious perceptions.

At this point we should note that it is not our intention to get involved in the lively discussion concerning relations and relational properties of the "vulgar individuals-David, Solomon, tables, chairs, etc." (to use Earman's expression; [30], p. 214) as opposed to relations and relational properties of the true individuals, i.e., monads. We also do not intend to discuss whether or not they are reducible to non-relational properties of such individuals. We nevertheless feel that we should briefly describe our position concerning the contrast between monadic relations and relations between "vulgar" individuals.

We think that "vulgar" individuals are only prima facie real and therefore in order to analyze relations holding between them we have to examine their individuality as phenomenal stemming from an intricate web of equally phenomenal relations holding between representors as representing one another. The relations between "vulgar" individuals then would be of second order. They would be considered as phenomenal bridges interconnecting groups of monads already phenomenally structured in a way responsible for their appearance as individuals. Such second-order relations can be analyzed in terms of first-order ones; they can be presented as infinite conjunctions of phenomenal relations obtaining between particular monads. Therefore it is enough to analyze only the particular elements of such infinite conjunctions. So we see our task as one of explaining phenomenal relations of first-order between two or more individual monads, in terms of their representational states as involving, among other things, indirect representations. We think that whenever Leibniz uses examples of relations between "vulgar" individuals for the purpose of explaining or, better, describing, his position concerning the reducibility of relations to the "modifications of the individual," he is doing it for the sake of finding a common ground for discussion with his readers. How else could he possibly talk about relations? He would most certainly not be able to find convincing examples, if he were to restrict himself to talking about relations between individual monads. In addition, using especially the example of David and Solomon, he gives the opportunity to his interpreter to explain such a second-order relation mainly via the modifications of the dominant monads (souls) of David and Solomon respectively. That is to say Leibniz's interpreter would be able to reduce directly a second-order relation between two "vulgar" individuals to a basic phenomenal one between two dominant monads. Such a relation would be a phenomenon well based upon the reality of monadic predicates of a representational nature, inherent in the two basic monads.

The decision we took to discuss mainly dyadic relations is due to two factors. First, the examples discussed by Leibniz himself are of dyadic relations, and, second, the treatment of polyadic relations would follow exactly the same pattern that we started unfolding in the case of dyadic ones. Let us again consider the relational statement:

$$S_1 \text{ is } R \text{ to } S_2.$$

We can distinguish at least two possibilities concerning R. The relation signified by it can be either symmetric or non-symmetric[10]. Let us assume, first, that R is symmetric and additionally that there exists a third monad S_3 not identical with either S_1 or S_2. The truth of the statement "S_1 is R to S_2 " rests on facts of the form S_1 represents (in specific manner M) S_2, S_2 represents (in specific manner M) S_1, and S_3 represents together S_1 as representing (in specific manner M) S_2 and S_2 as representing (in specific manner M) S_1. The manner of representing M corresponds to R and can be cashed out in terms of a more or less complex representational structure, made out of a specific combination of esoteric characteristics of the representation such as directness, indirectness, distinctness, confusedness, minuteness, and insensibility.

If the relation signified by R is non-symmetric, the relational statement

[10] A subclass of the symmetric relations is the class of all equivalence relations. On the other hand, all the relations of strict order are non-symmetric.

$$S_1 \text{ is R to } S_2$$

has to be conjoined with the relational statement

$$S_2 \text{ is P to } S_1$$[11] .

The truth of the statement "S_1 is R to S_2" and "S_2 is P to S_1" rests on facts of the form S_1 represents (in specific manner M) S_2, S_2 represents (in specific manner N) S_1 where M and N correspond to R and P respectively.

Both, the described above David / Solomon paternity / sonship relation and the relation described by Leibniz in the following passage (taken from *The Leibniz-Clarke Correspondence,* Leibniz's "Fifth Paper") are examples of the second sort, i.e., non-symmetric relations:

> The ratio of proportion between two lines L and M may be conceived three several ways: as a ratio of the greater L to the lesser M, as a ratio of the lesser M to the greater L, and, lastly, as something abstracted from both, that is, the ratio between L and M without considering which is the antecedent or which the consequent which the subject and which the object ... In the first way of considering them, L the greater, in the second M the lesser, is the subject of that accident which philosophers call relation. But, which of them will be the subject in the third way of considering them? It cannot be said that both of them, L and M together, are the subject of such an accident for if so, we would have an accident in two subjects, with on leg in one and the other in the other which is contrary to the notion of accidents. (A, 71; G, VII, 401.)

According to Leibniz, the first "way" of "conceiving" the phenomenal non - symmetric relation obtaining between S_1 and S_2 is as resting upon a representational fact of the form S_1 represents (in specific manner M) S_2 and the second as resting upon a representational fact of the form S_2 represents (in specific manner N) S_1, where M and N correspond respectively to R and P. There is a third "way," of course, and this is to consider the relation holding between S_1 and S_2 "as something abstracted from both ... without considering which is the antecedent or which the consequent; which the subject and which the object." This third "way," which appears problematic to Leibniz, seems to be so because there is not a unique subject and the relation cannot be cashed out in terms of a predicative inherence. The situation is as if the relation keeps its fine, impartial, and precarious balance between its two endpoints; and Leibniz is quick to say that "the relation common to both is a merely mental

[11] An example of such a situation is given by the pair of relational statements: Solomon is the son of David. David is the father of Solomon.

thing whose basis is the modifications of the individuals." According to our interpretation this relation is indeed a "mental" thing in a quite specific sense. It is not something obtaining in the real world of monads in the form of a relational bridge existing over and above the self-sufficient and detached monadic representors. It nevertheless appears to us as a relational bridge just because we, playing the role of the representor S_3, represent, in the framework of a unified representation, both S_1 as representing (in manner M) S_2 and S_1 as representing (in manner N) S_1. In other words, according to the interpretation we propose, Leibniz uses the phrase a "merely mental thing[12]" because (a) he wants to emphasize that the impartiality with which we represent such a relation with respect to its endpoints-whenever we are not one of them-is responsible for our thinking of it as existing over and above those endpoints and (b) he wants to drive a wedge between him and those who think that relations are real, i.e., that they are true inhabitants of the external world. It is worth noting once again, that the external world for Leibniz is the world of all the monads which represent one another as they represent one another as they represent one another ... and so on *ad infinitum*.

It is fair to say that the basic ingredients for the interpretation we propose are given by the Leibnizian text itself. According to Leibniz "the relation common to both is a merely mental thing," but it is a mental thing "whose basis is the modifications of the individuals." So using the interpretational scheme we described above, we can say that representor S_3 represents S_1 and S_2 as if they were R-related (this is what the characterization of "mental" refers to), but, although there is no such relation in the formal world of monads, the representation of S_3 is not a dream-like thing; on the contrary, S_3's representation is based on the representations of S_1 and S_2 as representing each other in specific manners. And since S_3's representation is a faithful one, it is right to say that S_3's internal viewing of it as a specific relational fact concerning S_1 and S_2 is a correct viewing.

Such correctness is of course guaranteed by God[13], who created a world governed by a pre-established harmonious coordination of the representational states of the monads. He is the one who can see it all. Monads represent the world as it is, that is, they represent one another as they represent. Nevertheless, they can distinctly represent only a finite part of it. So although such a representation is correct since, in a sense, it faithfully corresponds to what is real, it is also confused as to its infinitely many details.

---

[12] "... rem mere mentalem ... " (L, 609; G, II, 486)

[13] Or by the metaphysician.

We can now step back for a moment and take stock of what we have so far argued. Considering the truthfulness of a relational statement we said, following Sellars' interpretation, that it rests on representational facts of a certain sort; and we described in a very general way these facts. In what follows we summarize our account and connect it with what we take to be the Leibnizian version of the Cartesian formal / objective reality representational scheme.

(i) To the *prima facie* real relation R, that is, to the phenomenal relation R, corresponds a manner of representation M which can be cashed out in terms of a more or less complex representational structure of a representational whole existing in each particular representor with the representors S_1 and S_2 most prominent among them. The degree of complexity of such a structure corresponds to a similar degree of complexity of the structure of the phenomenal relation R. If, for example, R is the relation of *sonship*, then the *manner* of representation M which corresponds to it has to involve all sorts of representational characteristics having to do, first, with representing synchronic and diachronic configurations of monads representing all sorts of events constitutive of the sonship relation. For Solomon to be the son of David some prima facie real (i.e., phenomenal) relations and properties of certain configurations of monads have to obtain. A series of events, with David's conception of Solomon and Solomon's birth prominent among them, has to be involved in the analysis of this sonship relation. But, since every event appears as such by being grounded only on facts of a representational nature, in the last analysis to each event taking place in the phenomenal realm there corresponds a state or series of states of the world as it *really* is; namely, a world of monads in specific representational states tied together in a harmonious and pre-established way by the Creator.

So the sonship of Solomon has to be analyzed in terms of its simple phenomenal constituent parts. These simple phenomenal constituent parts have to be cashed out in terms of specific representational states of the world of monads. Finally, the truthfulness of the relational statement "Solomon is the son of David" can be thought of as resting on representational facts of the sort:

Solomon represents (in specific manner M) David.

The specificity of manner M can be determined as that which makes the representing of David by Solomon a complex representational structure, consisting of synchronic representations in Solomon of all of the specific representational states of the world of monads corresponding to the simple constituent parts of the phenomenal relation of his sonship to David and their interconnections. It is fair to say that such a complex representational structure concerning Solomon's sonship relation to David is inherent in Solomon in a unified way. It is also fair to say that the specificity of manner M is dependent not only upon the complexity of the particular representational structure itself (which could be distinctly seen in its infinite detailed entirety only by God), but also upon characteristics which concern the internal viewing of such a structure by the corresponding representor himself. Such characteristics are, e.g., indirectness, confusedness, minuteness, or even insensibility, either of the representation as a whole or of its parts

(ii) In the previous Section we insisted that in the Leibnizian scheme of things there is a domain of representables as "thought of" by God and ontologically dependent upon him. We also proposed that although Leibniz nowhere explicitly talks about his system by way of using the Cartesian dichotomy of formal versus objective reality, there is a sense in which such a dichotomy is present in his system[14]. It is true, of course, that there are substantial differences between the two systems, one of which concerns the fact that spatio-temporality is a characteristic of the formal in the Cartesian system and a characteristic of the objective in the Leibnizian. Such differences do not concern us here. What is important for the scope of this work is the formal/objective dichotomy as it appears in the Leibnizian system.

Having reminded ourselves of these points, we can once again consider the relational statement

$$S_1 \text{ is R to } S_2.$$

We have said already that the truth of the statement rests on facts of the form S_1 represents (in specific manner M) S_2, S_2 represents (in specific manner M) S_1, and S_3 represents together S_1 as representing (in specific manner M) S_2, and S_2 as representing (in specific manner M) S_1, provided that the *prima facie* real relation signified by R is a symmetric one[15]. Pushing our analysis a bit further we can say that S_3 represents both S_1 and S_2 in a unified representational act (in specific manner L), where the manner of representation L corresponds to R and incorporates all the specifics of the above description. By that we simply mean that S_1 and S_2 are parts of a unified representing,

[14] We additionally insisted that representables are ontologically prior to their realizations, namely, to their formal and objective counterparts. They are the ultimate points of reference, the metaphysical ultimate causes, or be-causes, for their correct correspondences.

[15] Our treatment of a non-symmetric relation follows a similar pattern, as we saw.

inherent in S_3, by way of being *objectively* or representationally present in this representing. Furthermore, we mean that the unified representing is a representing of both S_1 and S_2 as representing (in specific manner M) each other, where M already has the representational characteristics which correspond to the phenomenal characteristics of the *prima facie* real relation signified by R.

In order to make clear what we mean by the formal/objective dichotomy as it appears in the Leibnizian system, it is instructive to consider, e.g., the monad S_3 as being in the representational state we described above. *Objective reality* is then the phenomenal reality of the internal viewing of a particular representation, such as that which S_3 has of S_1 and S_2 as interconnected (in manner L) in the framework provided by the structural features of the representation itself. S_3 sees, in other words, S_1 and S_2 as R-related via an internal viewing or grasping of such a representation. This is an example of a model of how S_3 comes to believe that the phenomenal relation signified by R obtains between S_1 and S_2. For such a belief to be true it is necessary that it corresponds to the formal reality of the world of monads as they represent[16]. *Formal reality* in this case would be the reality of all other monads as they represent S_1 and S_2, as R-related, with monads S_1 and S_2 in the representational states we described above, prominent among these monads.

C. INDIRECT REPRESENTATIONS

If what really is in the world is only monads, i.e., substantial undivided units in their different states[17], then the question arises, as we have already said, as to what sort of thing these monadic states are supposed to be. The Leibnizian answer to it is straightforward. Monadic states are representational states; they are states of simple, i.e., partless representors representing the world from their point of view. Monads stripped of their states are, so to speak, undifferentiated, empty shells. Bare monads do not exist. Their only mode of existence is by way of being in those particular representational states which are always on the verge of changing. They express or represent in a way that both accounts for

---

[16] It is, of course, correct to say that no monad has direct epistemological access to the world of all other monads as they represent. Therefore, no knower can be sure about such a correspondence. On the other hand, beliefs about the external world can be justified by using criteria provided by the coherence of the world of real phenomena as opposed to imaginary ones. Leibniz's work *On the Method of Distinguishing Real from Imaginary Phenomena* (L, 336-366; G, VII, 319-322) is instructive and illuminating on this point.

[17] See, e.g., G, VI, 598; G, II, 119.

their specificity[18] and their mutual sharing of the world that they represent. But this world can only consist of monads being in particular and ever changing representational states. Therefore what each monad represents is all other monads as they represent. This situation forces us to admit that indirect representation is a dominant feature of the Leibnizian representational metaphysics. As a matter of fact, we have already used this idea. In our treatment of relations in the previous Section, we tacitly assumed that monads represent other monads as representing other monads and so on. We did that in order to make sense of the Leibnizian contention that although the relations appear to be real inhabitants of the world, what really is in the world is self-sufficient, separate, and non-interacting monads in ever changing specific representational states. Relations are only *prima facie* real. In fact they are phenomenal, well-founded and reducible to monadic predicates inherent, the way we described in the previous Section, in the monads involved.

We say that a monad A *directly represents* a monad or assemblage of monads B, or an unspecified something X, if the representation of B or X in A is not via a representation in A of a monad or assemblage of monads, as representing B or X. Accordingly, we say that a monad A *indirectly represents* a monad B or, more generally, an unspecified something X, if the representation of B or X in A is via a representation in A of a monad or assemblage of monads C as representing B or X. We can now define the notion of an indirect representation of a certain *degree* as follows: a direct representation can be thought of as a special case of indirect representation, namely, an indirect representation of degree equal to zero. An indirect representation of a monad B or of an assemblage of monads B or of an unspecified something X by a monad A is of the n-th degree if A represents a monad or assemblage of monads A_1 as representing a monad or assemblage of monads A_2 as representing ... as representing a monad or assemblage of monads A_n as representing B or X.

We can now give an example which shows how the notion of indirect representation can be used in order to explain what it could mean to say that a monad A represents monad B as *linearly nearer*[19] to A than monad C. A

---

[18] The spatial metaphor of a perspective representation is the key to what such a specificity amounts to.

[19] By "linearly nearer" we mean that monads A, B, and C are representing one another and they are represented by others as if they were positioned in the same straight line, with B positioned between A and C. At this point we will pretend we have discussed the problem of three monads being represented as co-linear and that we have arrived at a reasonable solution, permissible within the Leibnizian framework. In fact we will be discussing this problem in chapter III, where we will come to define the notion of a perspective line and propose a model for the

possible, simple model of the above phenomenal relation in terms of indirect representation could be as follows: A represents B as representing C. Before we continue our discussion of this example, we have to make two important remarks in anticipation of questions which might arise at this point.

1. Given the very important Leibnizian principle of continuity, a direct representation of a monad or a set of monads B by a monad A not only seems but is impossible. Since density, as a property of the phenomenal spatio-temporally extended is a constitutive part of what it is to be continuous, when I represent a monad B I have to represent it always as that which has *objective*[20] being in a more complex representing that I have, which involves in its turn a representation of B by some other monad C. In other words, since monads, because of density, do not phenomenally appear as strictly next to one another, when I represent a monad A as somewhere in space I do so by representing an infinitude of other monads as representing A. If we are to explain the phenomenal relation "linearly nearer" through indirect representation, there is no way we can hold that we can represent a monad A as absolutely *next* to us. In other words, if directly representing a monad A amounts to representing it as absolutely next to us, there can be no direct representation of a monad A by a monad B.

2. An indirect representation by a monad A, when considered as the metaphysical correlate of the phenomenally extended, involves always an infinite sequence of representations in the following sense: A represents A_1 as representing A_2 as representing A_3 ... and so on *ad infinitum*. This is so because of the principle of continuity and its correlate the, also Leibnizian, principle of plenitude. That is to say, the appearance of being extended involves density, which means, among other things, that there can be no spatial (i.e representational) vacuum of monadic representors. Furthermore, problematic is not only the infinitude of the set of terms of such a sequence when compared with the finite ability of the representor A to recognize these terms distinctly, but also the fact that at every point of the sequence – i.e., at the point where A_n represents A_{n+1}, where n is a natural number – the same problem as the one described in the first remark makes its appearance. Nevertheless, it is possible to make sense of the notion of an indirect representation of a finite degree for the case of the phenomenally spatio-temporally extended, using the contrast between distinct and confused perceptions. That is, we can talk about indirect representation using the spatial metaphor of n equal billiard balls perceived as

correspondence between the phenomenal relation of "spatially between" and its metaphysical counterpart.

[20] To use Cartesian terminology once again.

positioned in space so that their centers are co-linear, where n is a natural number greater than one. In such a case our perceptions of the billiard balls would be distinct as contrasted with our perceptions of what is in between them.

We can now return to our initial example of indirect representation, and attempt to present an explanatory model for the case of the phenomenal ternary relation expressed by the relational statement:

B is linearly nearer to A than C is.

As it has been mentioned, the above statement refers to a phenomenal linear ordering of A, B, and C. For such a statement to be true, certain facts of a representational nature have to obtain. The first is that monad A represents monad B as representing monad C; the second is that monad C represents monad B as representing monad A. A third could be of a negative nature, namely, that monad B does not represent monad A as directly representing C and that monad B does not represent monad C as directly representing monad A. A fourth monad D would then view such a phenomenal relation by incorporating in a unified representation inherent in D monad A as representing monad B as representing monad C, and monad C as representing monad B as representing monad A. Each one of the above representations are of an indirect nature. For instance, the representation of C by A and of A by C are of the first degree. The representation by D, on the other hand, of both C and A is of the second degree.

Before examining Leibniz's ideas of confused, petite, and unconscious perceptions, and as a prelude to it, we shall have to discuss a certain aspect of indirect representation which appears to have a rather paradoxical character. It involves, in a certain sense, an infinite regress. To make things as clear as possible, we have to examine a very simple example involving only two representors A and B. Let us assume that A directly represents B. Then since B represents the world and the world includes A as representing B, B has to represent A as representing B. But then, for the same reason, and in a similar manner, A has to represent B as representing A as representing B. This kind of reciprocal representation involves at every step a new representational item. So again B has to represent A as representing B as representing A as representing B and so on *ad infinitum*. This artificial example poses a rather serious problem for Leibniz's scheme of things. He himself seems to be worried, at least when he talks about self-reflection or apperception. Trying to prove that it is impossible to have a complete and distinct awareness of each one of the infinitely many constituents of a representation and at the same time trying to maintain the view that there are things we can represent in an unconscious and confused way, he states in the *New Essays on Human Understanding* that :

> ... it is impossible that we should always reflect explicitly on all our thoughts; for if we did, the mind would reflect on each reflection *ad infinitum*, without ever being able to move to a new thought. For example, in being aware of some present feeling, I should have always to think about that feeling, and further to think of thinking about it, and so on *ad infinitum*. It must be that I stop reflecting on all these reflections, and that eventually some thought is allowed to occur without being thought about; otherwise I would dwell for ever on the same thing. (NE, 118; G, V, 108.)

A few pages earlier he most explicitly proposes a way out of such a regress; it is a general way out, in fact, from the problem we are faced with, of a representor able to recognize only finitely many of an infinitude of constituents of some single representation that he has. We pay attention to only finitely many of these constituents because only finitely many can appear to us in a distinct way out of a mass of confused and unconscious ones. To Philalethes' (Locke's) statement "It is 'hard to conceive, that any thing should think, and not be conscious of it' " Theophilus (Leibniz) replies:

> That is undoubtedly the crux of the matter-the difficulty by which able people have been perplexed. But here is the way to escape from it. Bear in mind that we do think of many things all at once, but pay heed only to the thoughts that stand out most distinctly. That is inevitable; for if we were to take note of everything, we should have to direct our attention on an infinity of things at the same time ... (NE, 113; G, V, 103.)

Going back to our artificial example of the two representors A and B representing each other in a way that involves an infinite regress of indirect representations, we can use Leibniz's suggestion in order to break down such a regress. There are two basic constituents of the Leibnizian solution which deserve our attention. (a) Each representor has a representation of the other which in fact involves an infinite regress of indirect representations. But such a regress is not alarming because (b) each one of the representors represents distinctly only finitely many terms of such a regress, all the other terms being represented, but in a confused or even unconscious way.

In the more interesting case of the infinite regress of self-reflection the same solution applies. Since I am in the formal world in the state of reflecting on my reflections I have to represent together with the rest of the world myself as reflecting on my reflections, which could mean that I reflect on my reflecting on my reflections, and so on *ad infinitum*. Yet I stop reflecting on my reflections by ceasing to have new higher and higher reflections. That is to say,

I stop reflecting because "some thought is allowed to occur without being thought about," which means that an infinite regress of so-called self-reflections always exists, but only finitely many steps of it are distinct and so deserving of being called self-reflective with all the others being confused and insensible. *It is in a sense, as if there were representations in me of self-awareness of higher and higher order, which I am not aware of.*

D. PERCEPTION

For Leibniz, the real world, i.e., formal reality as opposed to the phenomenal, i.e. objective reality, consists of an infinitude of monads representing one another as representing one another. In the *Correspondence with John Bernoulli*, a monad is defined as "a substance which is truly one, i.e., not an aggregate of substances[21]". In the *Letters to De Volder* a monad is "a complete, simple substance[22]." In the *Letters to Des Bosses* "a perfect substance[23]" and in *The Monadology* "a simple substance," i.e., one "without parts" and with "no windows through which anything could enter or depart[24]." Monads, in short, are defined by Leibniz as substantial, undivided, partless, self-sufficient, windowless units. They all have the ability to represent and also to change, unconsciously and internally, moving from present representational states to future ones. In a letter to De Volder, dated June 30, 1704, Leibniz insists that:

> Indeed, considering the matter carefully, it may be said that there is nothing in the world except simple substances and in them perception and appetite. (L, 537; G, II, 270.)

But if what characterizes them is perception and appetite, what differentiates them must be, on the one hand, internal qualities, and, on the other, unconscious tendencies from one perception to another or, if you please, unconscious actions as individualized. Leibniz in *The Principles of Nature and of Grace Based on Reason* states that:

> ... one monad by itself and at a single moment cannot be distinguished from another except by its internal qualities and actions and these can only be its *perceptions*-that is to say the representations of the compound, or of that which is without, in the simple-and its *appetitions*-that is to say, its tendencies, from one perception to another-which are the principles of change. (L, 636; G, VI, 598.)

---

[21] GM, III, 537.
[22] G, II, 252.
[23] G, II, 306.

Furthermore, all monads "reach out to infinity or to the whole" in a confused way. They are limited because only a finite part of what they perceive can be perceived distinctly. Additionally they are differentiated "in the degrees of their distinct perceptions" as the following passage from *The Monadology* tells us:

> It is not in the objects represented that the monads are limited, but in the modification of their knowledge of the object. In a confused way they all reach out to infinity or to the whole, but are limited and differentiated in the degrees of their distinct perceptions. (L, 649; G, VI, 617.)

Leibniz, in another place (in a letter to Nicolas Remond, dated February 11, 1715), pointing out another important feature of monadic differentiation, epigrammatically states: "each monad is a living mirror of the universe according to its point of view" (L, 659; G, III, 636.) The characteristics that differentiate monads, apart from their individual internal unconscious activity, can be summarized in two basic categories: (a) characteristics having to do with perspective perception and (b) characteristics relevant to the distinction between confusedness and distinctness of perceptions as wholes or even confusedness and distinctness as contrasted in the framework of one and the same perception. Rescher adopts this categorization and uses the nice analogy of painting two different depictions of the same scene in order to present Leibniz's view:

> Since all monads are fundamentally alike in that every monad perceives everything else in the universe, a question can be raised as to how they can possibly differ from one another. Leibniz seems to think in terms of the analogy of painting, of two different depictions of exactly the same scene. Monads, he says, differ from one another not in what they perceive, but (1) in *point of view*, i.e., with differing features of the same thing they perceive; and (2) in *clearness of perception*, i.e., with differing faithfulness of representation of the various aspects of things[25].

Before we endeavor to examine the Leibnizian notions of confused and distinct perceptions, as they are used to explain certain aspects of both his metaphysics and his epistemology, it seems appropriate that we should clear the ground by determining what the term "perception" stands for in his system. According to at least one definition that he gives in a letter to Bourguet, dated December 1714), perception is "the representation of the plurality in the simple" (G, III, 574-575.) This is not the only place that he defines perception

---

[24] G, VI, 607.

[25] [125], p.71.

in those terms[26]. He additionally insists that "there is nothing in the world except simple substances and in them perception and appetite." So, since the world, according to Leibniz, really consists of monads perceiving all the other monads, themselves included, as being in certain perceptual states, and since all there is in a monad is its perceptions and appetitions, "perceiving" is the sort of term, in the Leibnizian terminology, signifying any kind of representational act taking place in a monad, abstract thought included. Perception, conceived of as the representation or the expression of the many in the one, includes everything from completely unconscious representational states of a monad, representational states which could be classified as sensations, to representations corresponding to what Leibniz calls apperception.

Leibniz is not quite consistent in his usage of the term "perception." As we already said, "perception" is mainly used as an all-inclusive term signifying any sort of mental state of a monad. There are passages, however, where it seems to be used otherwise. In the following passage, e.g., from a letter to Rudolph Christian Wagner, dated June 4, 1710, perception is distinguished from sensation .

> ... I could not have the sensation of green unless I perceived the blue and yellow from which it results. At the same time I do not have the sensation of blue and yellow unless a microscope is used. (W, 505; G, VII, 529.)

In the above passage, Leibniz seems to reserve the term "perception" for the signification of minute constituent parts of a more complex representation, which by themselves happen to be insensible and which, by their confused interweaving give rise to the sensation of green, as a novel emergent. Most of the time, of course, he uses the term "perception" as opposed to the term "sensation" in a consciously restricted sense. In the following passage from a letter of his to Des Maizeaux dated July 8, 1711, Leibniz seems to call "perceptual" the representational states of simple monads constitutive of organic bodies which do not belong to animals, as opposed to the more distinct representational states of the dominant monad of an animal, which states he characterizes by the term "sensation." In the last sentence of the same passage, however, he seems to ignore the previous usage by insisting that there is "an infinity of degrees in perception," tacitly implying that the term "sensation" characterizes a certain segment of a continuous spectrum of degrees of one and the same thing, namely, degrees of distinctness or confusedness of perception:

[26] See, e.g., G, II, 121; G, III, 69; G, VI, 598, 608; G, VII, 529.

> But if there are in Nature other living organic bodies besides animals ... these bodies also have their simple substances or *monads*, which give them light, that is to say, perception and appetite, although it is not necessary that this perception be a sensation. There is apparently an infinity of degrees in perception, and consequently in *living things*. (G, VII, 535.)

A situation similar to the one described above seems to obtain, if one considers Leibniz's use of the term "perception" as opposed to the term "apperception." Let us consider the following passage from *The Monadology*:

> 14. The passing state which enfolds and represents a multitude in unity or in the simple substance is merely what is called *perception*. This must be distinguished from apperception or from consciousness, as what follows will make clear. It is in this that the Cartesians made a great mistake, for they disregarded perceptions which are not perceived. It is this, too, which led them to believe that only spirits are monads and that there are no souls in beasts or other entelechies. It led them into the popular confusion of a long stupor with death in a rigorous sense, which made them support the Scholastic prejudice that souls are entirely separate, and even confirmed some ill - balanced minds in a belief in the mortality of the soul. (L, 644; G, VI, 608-609.)

In this passage Leibniz seems to be sharply distinguishing perception from apperception. It is nonetheless fair to say that as he goes on talking about the "great mistake" of the Cartesians, he seems to have in mind perception as an all embracing term, including apperception also. If the Cartesians "disregarded perceptions which are not perceived" they must not have done so concerning perceptions which are perceived. "It is this ... which led them to believe that only spirits are monads and that there are no souls in beasts or other entelechies." The picture that Leibniz draws of the Cartesians is basically correct. We can therefore say, using his terminology, that what the Cartesians believed was that only spirits and souls are capable of conscious perceiving-conscious perceiving being the only way of perceiving for them[27]. The interesting point for us, however, is not what Leibniz thought and believed about the Cartesians, but that in talking about them using his own terminology, he implicitly assumes that all forms of mental activity-unconscious perceiving included-are forms of perception. Let us consider one more passage from the *New Essays on Human Understanding* supportive of the thesis that apperception, although ostensibly sharply distinguished from perception by Leibniz can be thought of as a special kind of perception.

---

[27] That is, the Cartesians believed conscious perceiving to be the only form of perceiving.

> ... we always have an infinity of minute perceptions without perceiving them. We are never without perceptions but it is necessary we are often without *apperceptions* namely when they are not perceptions which are noticed by being distinct. (NE, 161-162; G, V, 148.)

So "we are never without perceptions," but we can be "without *apperceptions*." Whenever, on the other hand, we apperceive, we simply perceive our perceptions, i.e., we notice them or, better, we notice finitely many parts of them, those which are appropriately distinct. Such distinctness in the case of apperception is, according to Leibniz, due to our being conscious of having those perceptions. It is due to the fact that together with our perceiving of them we are aware of ourselves as perceiving them. Apperception is a special species of indirect representation. Apperceiving something amounts to representing something together with representing ourselves as representing this something. So we can consider apperception as a special case of perception, plus self-awareness or self-reflection. Self-awareness is the perception of the "I" as a perceiver. Leibniz hesitates to call perception such a self-awareness. Yet, we should not consider that as a piece of conclusive evidence for the separate classification of perception and apperception. Since all there is in a monad is its perceptions and its appetition, it seems to us that we should classify "apperception" as a special kind of term in the Leibnizian terminology, meant to signify the upper part of the continuous spectrum of perceptions as they are ordered according to their different degrees of distinctness or confusedness.

We argued that Leibniz is not very careful in his usage of the term "perception." He fluctuates between characterizing it as an all-inclusive term signifying all forms of monadic representation and as a term meant to signify only lower forms of such a representation or expression. Sometimes, when he is more careful, he uses an appropriate adjective to serve as a meaning qualifier for it. The following passage, from a letter of Leibniz to Antoine Arnauld, dated 9 October, 1687, is quite instructive as an example of such a case. In it, Leibniz talks about expression (i.e. representation), a species of which is "natural perception" in a single monad. At the same time he talks about "animal sensation" and "intellectual knowledge", the latter being what one gets through apperception, i.e. through perceiving oneself as perceiving.

> One thing expresses another (in my terminology) when there is a constant and regular relationship between what can be said of one and of the other. It is in this way that a projection in perspective expresses a geometric figure. Expression is common to all the forms and is a genus of which natural perception, animal sensation and intellectual knowledge are species. In natural perception and sensation it suffices that what is divisible and material

> and is found dispersed among several beings should be expressed or represented in a single indivisible being or in a substance which is endowed with a true unity. The possibility of such a representation of several things in one cannot be doubted, since our soul provides us with an example of it. But in the rational soul this representation is accompanied by consciousness and it is then that it is called thought. (M, 144; G, II, 112.)

Some remarks concerning this quotation are in order. (a) Leibniz, in using the composite term "natural perception" seems to be referring to the sort of perception that monads all the way down the ladder of being always have. More generally, he is referring to perceptions which can pass completely unnoticed, i.e., perceptions which can be called unconscious or insensible. No part of them is distinct and even souls can sometimes have them, when they are, for example, in a state of lethargy[28]. (b) Animal sensation, as the second species of monadic expression or representation, is characterized by some degree of distinctness of representation, which is responsible for the memory of the sensation that follows[29]. (c) Intellectual knowledge as the third species of monadic expression or representation is what characterizes rational souls. The degree of distinctness of the representation is extremely high. Intellectual knowledge is separated from sensations by the fact that it belongs to the area of representations characterized by the awareness of the "I," i.e., the self-awareness of the representor as he represents; it belongs to the area where thought is possible, the area, in other words, of apperception.

In view of all of the above, and by way of summarizing the basic position we argued for in this Section, we will henceforth consider perception as synonymous with monadic representation or expression; one should keep in mind that, whenever appropriate, we shall be using more specific Leibnizian terms in characterizing different kinds of perceptions. In other words, the view that we will adopt in what follows is that every monadic representational act is a perceptual act and vice-versa. Monadic perceptions or representations, then, will be considered as graded according to their distinctness or confusedness as wholes and according to the distinctness or confusedness of their parts. They will be considered as if their grading forms a continuous spectrum[30], which resembles a closed straight line segment, the topmost point of which would correspond to an ideal of absolute distinctness for perception considered both

---

[28] NE, 139; G, V, 127.

[29] G, VI, 599.

[30] A view that ties up with Leibniz's doctrine that every gradation of beings has to be continuous. For the correspondence between "the degrees in perception" and the gradation of "*living things*" see again G, VII, 535.

as a whole and as a manifold of its infinitely many parts; this is an ideal that could be attained only by the Supreme Being, if the word "perception," with all of its anthropomorphic imperfection could be used in this case. The lowest point of such a line segment would refer to perceptions absolutely confused and unconscious; these are perceptions which pass completely unnoticed by the perceiver, without leaving any traces in his memory. If such traces were to exist they would necessarily be of a representational nature.

The continuous spectrum of the different degrees of perceptual distinctness, which as we have mentioned could be thought of as a closed straight line segment, would then be divided into three segments[31]. The lowest one would be the part of the spectrum referring to natural perception, the middle one to animal sensation, and the third one to apperception and intellectual knowledge. It is clear that the topmost segment would be, in reality, open-ended, because the point that closes it upwards refers to an ideal that no perceptual state of any created being could correspond to. We can even imagine the possibility of the existence of created beings who could have finitely many distinct perceptions at one particular time, in such a way that there was not a finite numerical upper bound to how many such distinct perceptions they could have. Yet the possibility of the existence of a created being, resembling God in his infinite capacity for distinct knowledge, would be out of the question. God for Leibniz-and for every 17th century Christian thinker, for that matter-is an approachable, but never reached upper limit.

E. DISTINCT, CONFUSED, PETITE, AND UNCONSCIOUS PERCEPTIONS

Perceptions, according to Leibniz, can be *distinct* or *confused*. Distinctness or confusedness are characteristics not only of a perception as a complete and unified representational state of a monad, but also characteristics of parts of it. A representational state of a monad is a perspective mirroring of the world, and as such consists of an infinitude of partial representations corresponding to the infinitude of the particularities of this world. The way we see the world is characterized by a distinct perceiving of certain parts of it and by a confused, and even unconscious perceiving of the rest.

As we have said, perceptions (and appetitions) are all there is in a Leibnizian monad. In the previous Section we proposed the thesis that the main usage of the term "perception" by Leibniz is as synonymous with "monadic

---

[31] In fact, four, as we shall see in the next section. That is if we consider perceptions of dominant monads of plants as positioned in between sensations of animals and completely unconscious perceptions.

representation or expression." Distinctness and confusedness of perceptions are characteristics which are amenable to a continuous grading. Such a grading leads to a parallel grading of perceptions and consequently to a continuous grading of monads themselves.

Absolute confusedness of their perceptions is a characteristic of primitive entelechies or simple monads. A primitive entelechy or simple monad, as a perceiver, does not *perceive* its perceptions, which means that its perceptions are so confused that the monad is not only unconscious of parts of them, but also of them as wholes. This is not to say that monads higher up in the continuous ladder of being do not sometimes also have partly or completely unconscious perceptions, perceptions that is, which do not *perceive*. Dreamless sleep, lethargy, death are not cases of absence of any perceptions. On the contrary, the continuous unfolding of perceptions never stops. Perceptions continue flowing, but the perceiver is unconscious of them. Neither they as wholes nor parts of them are sufficiently distinct to be noticed. Such a lack of distinctness leads to a kind of amnesia on the part of the perceiver. By "amnesia" we do not mean that no traces of his perceptions are left in him. On the contrary, we mean that such traces remain unconscious in him. Though forgotten, some day their turn will come to contribute to some noticeable perception. The following passage from the *New Essays on Human Understanding* is quite illuminating in this respect:

> In man's case ... perceptions are accompanied by the power to reflect, which turns into actual reflection when there are the means for it. But when a man is reduced to a state where it is as though he were in a lethargy, and where he has almost no feeling, he does lose reflection and awareness ... Nevertheless his faculties and dispositions, both innate and acquired and even the impressions he receives in this state of confusion, still continue; they are not obliterated though they are forgotten. Some day, their turn will come to contribute to some noticeable result; for nothing in nature is useless, all confusion must be resolved, and even the animals which have sunk into a condition of stupidity, must return at last to perceptions of a higher degree. Since simple substances endure for ever it is wrong to judge of eternity from a few years. (NE, 139; G, V, 127.)

That primitive entelechies or simple substances exist should not be thought of as contradicting the basic Leibnizian dogma, that nature is not only everywhere animated, but also everywhere organic. It is rather that a primitive entelechy is a monad whose perceptions are completely confused for long periods of time, as, for example, before the birth and after the death of the organism which corresponds to it, and of which the primitive entelechy existing before and after the organism would be the dominant monad. Nature is everywhere organic in the sense that monads are going in a continuous way through phases of representational states which, by their differing degrees of distinctness or confusedness, are metaphysically responsible for the relegation of these monads into the state of a primitive entelechy, or for their advancement to the state of a being higher up in the ladder. In other words, in the complete concept of a monad is included as a blueprint the fact of being organismic, cashed out in terms of its internal representational structure. This structure, together with the representational structures of all other monads in the world and, more specifically, with the ones that correspond to what we would call its "body," give rise to the well-founded phenomenon of a mortal organism; the dominant monad of the mortal organism is our initial one. The following long quotation from a letter of Leibniz to Wagner, dated June 4, 1710, is quite illuminating:

> As ... mind is rational soul, so soul is sentient life, and life is perceptive principle ... according to my opinion, not only are all lives, all souls, all minds, all primitive entelechies, everlasting, but also that to each primitive entelechy or each vital principle there is perpetually united a certain natural mechanism, which comes to us under the name of organic body: which mechanism, moreover, even although it preserves its form in general, remains in flux, and is, like the ship of Theseus, perpetually repaired. Nor, therefore, can we be certain that the smallest particle of matter received by us at birth, remains in our body, even although the same mechanism is by degrees completely transformed, augmented, diminished, involved or evolved... Every natural mechanism, therefore, has this quality, that it is never completely destructible, since, however thick a covering may be dissolved, there always remains a little mechanism not yet destroyed ... And we ought to be the less astonished at this for this reason, that nature is everywhere organic and ordered by a most wise author for certain ends. ...in truth, a soul or an animal before birth or after death differs from a soul or an animal living the present life only by conditions of things and degrees of perfections... (W, 505-506; G, VII, 529-530.)

So primitive entelechies (simple monads) not only do exist but,(a) are organic in a quite specific sense and (b) are completely unconscious perceivers of their perceptions as long as they are not in their prime, i.e., as long as being an organism remains only a potentiality in them. As absolute confusedness (and therefore unconsciousness) of their perceptions is a characteristic of simple monads, so absolute distinctness of his "perceptions" is a characteristic of God. In between and according to the degree of the distinctness of their perceptions are situated "vegetative, sensitive and reasonable souls" (G, VI, 521.) Sensation, according to Leibniz, is a kind of confused perceptual state which, nevertheless, considered as a whole, is distinct enough to be noticed by the perceiver. It is exactly what characterizes the dominant monads of animals. This is not to say that rational souls are not capable of having sensations. On the contrary, they are capable of having not only sensations, but also of having perceptions of a much higher order, i.e., perceptions of themselves as perceiving.

Sensations, because of their distinctness as perceptions, are followed by memory. According to Leibniz, (passage from *The Principles of Nature and of Grace Based on Reason*) the mechanism of sensing can be described as follows:

> Together with a particular body, each monad makes a living substance. Thus not only is there life everywhere, joined to members or organs, but there are also infinite degrees of it in the monads, some of which dominate more or less over others. But when the monad has organs so adjusted that by means of them the impressions which are received, and consequently also the perceptions which represent these impressions, are heightened and distinguished (as, for example, when rays of light are concentrated by means of the shape of the humors of the eye and act with greater force), then this may amount to *sentiment*, that is to say, to a perception accompanied by *memory*-a perception of which there remains a kind of echo for a long time, which makes itself heard on occasion. Such a living being is called an *animal*, as its monad is called a soul. (L, 637; G, VI, 599.)

A characteristic, therefore, of a soul is that it is capable of sensing. The capability of sensing, in turn, implies the capability of remembering what has been sensed. The capability of remembering involves a kind of consecutiveness as well as a comparison of representations of the past with representations of the present and representations (of expectations) of a possible future. This is how memory comes to stimulate reason. Reason is not a necessary consequence of memory. Nevertherless reason needs memory to get stimulated. In *The Monadology* Leibniz says:

> 26. Memory provides a kind of *consecutiveness* to souls which stimulates reason but which must be distinguished from it. Thus we see that when *animals* have a perception of something which strikes them and of which they have had a similar perception previously, they are led by the representation of their memory to expect whatever was connected with it in this earlier perception and so come to have sensations like those which they had before. When one shows a stick to dogs, for example, they remember the pain it has caused them and whine and run away. (L, 645; G, VI, 611.)

There are two important features of sensations to consider before we move on: (a) According to Leibniz a sensation is a distinct enough perception of a whole the ingredients of which are mixed together. These ingredients are *petite perceptions*, which, if they were to be distinctly perceived, would result in a different viewing of the whole. I do not have the sensation of green[32] unless I perceive the blue and yellow from which it results. So, in sensing green I perceive distinctly, yet as an undifferentiated whole, the confused mass of my petite perceptions of blue and yellow. That is, I perceive distinctly blue and yellow as intermingled without having at the same time a distinct perception of them as they are by themselves. (b) Sensation as a heightened confused perception would not be possible if there were no sense organs. Consequently, sensations could not be "experienced" by a monad, which would not be the soul of a well-formed animal. But since well-formed animals, as such, are not to be found in the world of monads, the vague Leibnizian contention should be phrased as follows: a monad would not be the soul of an animal if there were no other monads in representational states coordinated with the representational states of this soul-monad in the appropriate way. In other words, when the soul-monad senses, it simply represents these other monads, which are phenomenally constitutive of its sense organs, as receiving impressions, i.e., physical imprints by, objects such as e.g., light rays, sound waves, sticks, stones, etc. "Receiving impressions," finally, has to be cashed out in terms of appropriately coordinated representational states of the world of monads and especially of the monads immediately involved in this phenomenal interaction.

If sensation is that which characterizes animals, what about plants? Can we say that they also possess a soul? Can we attribute to them sensation? Leibniz seems to be terminologically inconsistent with respect to the second question. There are places where he uses the term "soul" in a broad sense, as for example, when he talks about "vegetative[33]" ones, and other places where he uses it in a restricted sense, referring only to the dominant monads of animals and rational beings[34].

Concerning the third question, the answer is most definitely a negative one. Plants do not have sensations. Yet, they do have something resembling sensation. But if plants are to be distinguished from animals by not having sensations, what about their hierarchical position with respect to the lowest end of the spectrum, i.e., with respect to primitive entelechies, such as those constitutive of a rock? Furthermore, is there a dominant monad for every

---

[32] See G, VII, 529.

[33] G, VI, 521.

[34] See, e.g., G, VI, 599-600.

plant? The following quotation from the *New Essays on Human Understanding* is quite illuminating:

> THEO. The great analogy which exists between plants and animals inclines me to believe that there is some perception and appetite even in plants; and if there is a vegetative soul, as is generally thought, then it must have perception ... I agree that the movements of what are called sensitive plants result from mechanism...
>
> PHIL. Indeed 'I cannot but think, there is some small dull perception' even in such animals as oysters and cockles. For 'quickness of sensation [would only] be an inconvenience to an animal that must lie still, where chance has once placed it; and there receive the afflux of colder or warmer, clean or foul water, as it happens to come to it.'
>
> THEO. Very good, and I believe that almost the same could be said about plants. (NE, 139; G, V, 126-127.)

In the following passage from a letter of Leibniz to Varignon (1702), it becomes apparent that the Leibnizian continuous spectrum of beings has as one of its segments that of the plants, which lies in between the segment of the animals and the segment of the fossils. These segments are linked together in such a way that, with the addition of the segments of human beings and of "those bodies which the senses and the imagination represent to us as perfectly dead and formless", form the linearly continuous spectrum of beings:

> Men are ... linked with the animals, these with the plants and the latter directly with the fossils, which in their turn are linked with those bodies which the senses and the imagination represent to us as perfectly dead and formless. (W, 187; BC, II, 558.)

In the continuous gradation of beings plants are placed below men and animals and above fossils and bodies (rocks, chairs, tables, etc.) "which the senses and the imagination represent to us as perfectly dead and formless." But since such a gradation corresponds to the gradation of perceptions with respect to their confusedness and distinctness, one has to admit that Leibniz considers the perceptions of the dominant monad of a plant-if such a thing exists-as, on the one hand, not absolutely confused and, on the other (though resembling sensations by being, so to speak, "dull" ones) as the sort of representational states which cannot be properly classified as sensations. To complete the description, as we have said, one has to consider also the perceptions of rational souls. Intellectual thought as based upon introspection-i.e., apperception, that is the perception of the "I" as perceiving - is what characterizes such souls. For the purposes of this section it is enough to say that "apperception" signifies the most distinct upper segment of the continuous spectrum of perceptual states of monads classified according to their degree of distinctness or confusedness.

Perceptions, according to Leibniz, are representations of the *formal* world of monads as they actually represent. The formal world of monads is not spatio-temporal. Monads are, nevertheless, represented by other monads as if they were the real inhabitants of such a world. On the other hand, they do not belong to the *objective* spatio-temporal reality of a representation, which means that they are not the spatio-temporal shadows of their own selves. We perceive the world of monads (as representing one another) as if space and time were absolute containers of things moving, changing, interacting. We are sometimes, of course, in representational states which, internally viewed, do not seem to constitute a direct spatio-temporal mirroring of formal reality. Such representational states are, for example, those that could be classified as corresponding to what we call abstract thought.

The world appears to us as spatio-temporal. We represent what there is, as forming extended wholes in a simultaneous orderly arrangement. However we never represent distinctly the ultimate constituent units of such wholes. By "ultimate constituent units" we mean particular monads as they are represented in us. We represent them, as units, confusedly and therefore insensibly. At the phenomenal level, i.e., at the level of a particular representation of the world, this leads, via the principle of continuity, to the contention that we can never arrive at distinct representations of particular monads as "living points, or points endowed with forms" by way of subdividing extended objects such as, e.g., "a piece of flint".[35] The number of such subdivisions performed by any finite perceiver would necessarily be finite and always leading to the partition of the piece of flint into extended parts, and not to its dissolution into distinct representations of substantial points. According to Leibniz, when we perceive an extended object as positioned in the simultaneous orderly arrangement of what we see as extended, we do *distinctly* perceive an item whose ultimate constituent units are *confusedly* represented. It is, in other words, not only that the formal world of monads is not spatial and temporal-which means that monads and their states are not located in space and time - but also that when monads are represented as located in space, they are represented confusedly as forming, together with an infinitude of others monads, extended material objects simultaneously arranged. Our representation of the world is a representation of an infinitude of monads as representing one another. We are finite representors not in the sense that we do not represent such an infinitude,

[35] GM, III, 552.

but in the sense that we can only have perceptions wherein, as it were, we can distinguish only finitely many representational items. Such items are not representations of isolated individual monads, but the resultants of confused masses of petite perceptions. According to McRae[36]:

> A confused mass of perceptions, taken in the mass, may be sufficiently forceful and distinguishable from other confused masses of perceptions to command attention. This confused mass of an infinitude of little perceptions will in that case form a distinct perception for the apperceiving[37] mind. It will be distinct in terms of its external relations to other perceptions from which it is distinguished. It will be confused in terms of the infinitude of little perceptions of which it is composed, these being too insignificant or too feeble to be separately apperceived.

Petite perceptions, though insensible, have an overall effect. The result is a distinct perception, which has built into it the confusion of its parts. Such a distinct perception is not the simple aggregate of its parts, but a novel emergent. Sensing green is the novel emergent of insensibly perceiving blue and yellow as if they were mixed together. According to Leibniz, if I were to use a microscope, I would be able to perceive distinctly all the little perceptions of yellow and of blue as separated from one another[38]. I would have, in other words, the view of an area covered with little spots of yellow and of blue next to one another. This is not to say that petite perceptions are restricted to what we would today call visual sense perceptions. On the contrary, Leibniz's main examples of how it is that petite perceptions contribute to the distinctness of a perceptible whole, confused in its details, are examples of auditory sense perceptions (G, VI, 534; G, VI, 604.) Additionally, whenever he talks about sensation in general he presupposes his analysis of a confused mass of perceptions into minute and indistinguishable parts.

Two questions arise at this point. What is it that makes petite perceptions insensible? And why can a microscope help us see finer details, thus making it possible to distinguish petite perceptions which were insensible to us before? Before answering these questions, it will be helpful to consider Leibniz's position concerning infinite divisibility of the extended. According to Leibniz,

---

[36] See [106], p. 36.

[37] The use of the notion of apperception here is rather unfortunate. After all animals can sense, i.e., they can have distinct enough perceptions of confused masses of petite ones. Animals, on the other hand, are not capable of apperceiving. Rational souls can also have distinct enough perceptions without having to be in the state of apperceiving them. This minor point aside, we have to agree with McRae's description of how we come to have distinct perceptions of confused masses of petite ones.

[38] See, again, G, VII, 529.

what is extended is not only infinitely divisible, but actually infinitely divided[39]. This is so because everything appearing as extended has as its metaphysical counterpart an infinitude of monads representing one another in a way that each one is represented as positioned in such an extended, massive, and continuous (i.e., infinitely divisible) whole. So what is metaphysically prior to a phenomenally extended whole is its monadic unextended units, as representing one another, and not its phenomenally extended parts. An extended whole is phenomenally infinitely divisible because it is already metaphysically divided into an infinite multitude of monads as they are represented. In the following passage from a letter of Leibniz to De Volder, dated June 30, 1704, one can see quite clearly that infinite divisibility of the phenomenally extended massive wholes is firmly based upon the non-spatial reality of the monadic representors as they do represent one another. Additionally, such divisibility can be carried out to infinity and phenomena can be divided into lesser and lesser phenomena distinguishable by more and more subtle animals:

> And granted that these divisions proceed to infinity, they are nevertheless all the results of fixed primary constituents or real unities (monads), though infinite in number. Accurately speaking, however, matter is *not composed* of these constitutive units but results from them, since mass is nothing but a phenomenon grounded in things, like the rainbow, or the mock-sun, and all reality belongs to unities. Phenomena can themselves always be divided into lesser phenomena which could be observed by other, more subtle animals, and we can never arrive at smallest phenomena. Substantial unities are *not parts*, but foundations of phenomena.. (L, 536; G, II, 268.)

In order to answer the questions raised in the last paragraph, it is necessary, in view of the above quoted passage, to explain more thoroughly what Leibniz means when he says that the extended is not simply infinitely divisible, but actually infinitely divided. What is extended is continuous and therefore everywhere dense. Being extended, on the other hand, is a phenomenal property of a whole, whose metaphysical foundation is an infinite multitude of monads representing one another in such a way that they appear to themselves, and to every other monad, as forming a continuous extended whole. If we want to understand what continuity and therefore density of the extended means in terms of the world of monads representing one another, we have to use the notion of indirect representation, that is, the notion of a monad A representing a monad B or, more generally, an unspecified something X via a representation in A, of a monad or assemblage of monads C as representing B or X.

---

[39] See, e.g., G, I, 416; G, VI, 618.

Monads are not in space, but they represent one another as holding spatial positions. If this is so, and since a monad as represented is not extended, when a monad represents another one it does it only, so to speak, through the eyes of "intermediate[40]" monads. Continuity of the phenomenally extended forces us to adopt the view that a monad *never*[41] directly represents another monad. As a matter of fact, since continuity of the phenomenally extended involves an infinitude of potential divisions between any other two such divisions, the metaphysical counterpart of it would be that a monad represents any other monad only through indirect representation involving infinitely many monads "in between." The simplest model for such a case would be that of an infinite sequence of representations of the form: monad A_1 represents monad A_2 as representing monad A_3 as representing ... and so on *ad infinitum*. But such a model would be too simple, since density would force us to accept that "in between" any two monads representing each other, or as represented as representing each other, there is an uncountable infinitude of such sequences. The only way out of the problem would be to adopt the already familiar Leibnizian solution: though no monad can represent distinctly such indirect representations of infinite length in detail, it can represent them as undifferentiated wholes which, by their mutual interweaving in one and the same perception, give rise to the phenomenally extended. If this is so, then we can distinguish only extended objects in a perception and, since we are finite representors, only finitely many of them.

The reason, then, that we cannot perceive any monad in isolation is that, due to the phenomenal continuity of the extended, every monad is so densely "crowded" by other monads that no division would enable us to reach it. It is as if the requirement of perceiving a monad in isolation were tantamount to creating a vacuum of monads all around it. Metaphysically speaking, it would be tantamount to asking for a direct representation of a monad by us, where directness could mean an immediate and distinct representation of it in us

---

40 In chapter III, we will discuss more thoroughly what such an "intermediacy" amounts to.

41 It is true, of course, that when I see the wall in front of me in my room, I represent the monads constitutive of the "surface" of the wall in a "direct" way. But the phenomenal world is a plenum. "Between" me and the wall there is an infinitude of monads as represented by me. Therefore I represent the wall through, as it were, the eyes of these intermediate monads. Why then do I not represent these intermediate monads? I do represent them, but unconsciously. After all, I represent the whole universe in its minutest detail. But why am I unconscious of monads "nearer" to me than the wall? Since such a phenomenal transparency would have an objective feature, in the sense that not only I, but everyone would be unconscious of the monads "nearer" to me than the wall-if he were to stand where I stand-an explanation of such phenomena would have to be cashed out in terms of characteristics of the representational structures of these "intermediate" monads themselves as causes of my unconscious representations of them.

without the web of indirect representations (confused in its details and distinct as a whole) which in any case would be involved. Density, as a property of the extended, and the phenomenal unextendedness of a monad as it is represented, are responsible for its non-distinct representation by any other representor.

At this point we can ask again the question: what is it that makes petite perceptions insensible to the perceiver? If by "petite perceptions"[42] we mean the unconscious perceptions we have of individual monads, then they are insensible for the reasons we gave above. But this is not what Leibniz means when he uses the term. According to him "phenomena can always be divided into lesser phenomena which could be observed by other, more subtle animals and we can never arrive at smallest phenomena." What Leibniz tells us here can be analyzed into the following components:

(1) "Substantial units are not *parts*, but foundations of the phenomena." First, by themselves they are not phenomenal, but real unities and, second, they are not represented by other representors as extended. If they were, they would be phenomenally divisible and this would contradict the basic dogma of Leibniz's representational metaphysics, according to which what is represented as having parts cannot be *really* (i.e., metaphysically) partless. Substantial units, as represented, are not parts of the extended, because every partition of it is a phenomenal subdivision of it into smaller parts, also extended[43].

(2) "Phenomena can always be divided into lesser phenomena which could be observed by other, more subtle animals..." This is an important statement of Leibniz because it implicitly shows us the way to answer the initial question. We can distinguish two components in it. First, the statement that "phenomena can always be divided into lesser phenomena," applied to the case of what appears as extended, means that there is no spatially extended item which could be indivisible either actually or in principle. If such an item could exist as a representation of something, this something would have to be an indivisible, metaphysical, i.e., real unit. It would have to be the representation of a monad because if it were not, it would have to be the representation of a multitude of monads. But a multitude of monads is already an actually divided artificial entity. So if we were to adopt the second alternative, we would again be led to a conclusion contradicting a basic Leibnizian dogma, according to which what is *really* a plurality cannot be represented as an individual unit. If, on the other hand, we were to adopt the first alternative, we would be in an equally difficult position. Monads cannot be represented as indivisible

[42] We still consider here visual petite perceptions as our paradigm.

[43] In other words, *parts* of the extended can only be *extended.*

extended wholes because that would result in a phenomenal world of extended atoms, contradicting one of the basic principles of the Leibnizian system, namely, his principle of continuity which implies density and therefore infinite divisibility of the extended. Secondly, since the subdivision of phenomena into lesser and lesser ones could be paired with the existence of more and more subtle animals able to perceive them, we are led to a hidden Leibnizian premise, that for a particular monad there must be a characteristic of its representational structure concerning its inability to represent extended things of magnitude less than a certain one. If this is not so, why does Leibniz need to say that the subdivision of phenomena into lesser and lesser ones could be paired with the existence of subtler and subtler animals able to perceive them? In saying it he presupposes that a given monad cannot distinctly represent all possible subdivisions of phenomena *ad infinitum*. On the other hand, he seems to think that the *continuous* subdivision of phenomena into lesser ones is paired with a *continuous* chain of more and more subtle beings, which can observe them. It is quite interesting to note that when he talks about sensing green he insists that we would not be able to represent distinctly blue and yellow from which the green results; we would need a microscope to do it. So we are led to the idea that there is a certain boundary, a *minimum perceptibile*, so to speak, beyond which the monad cannot distinguish finer perceptual details. Such a boundary depends on the monad and more specifically on the monad's particular representational structure, being not an absolute minimum perceptible but a relative one. We can say therefore that petite perceptions are insensible or unconscious because they are not big enough to exceed the limitation imposed by such a boundary. This is exactly what Russell has in mind when he says:

> Again, as regards minute perceptions, Leibniz holds, with modern psychophysics, that a perception must reach a certain magnitude before we become aware of it, and thus sufficient minute perceptions are necessarily unconscious[44].

We should make clear that for Leibniz the idea of minute or petite perceptions and its correlate, the minimum perceptibile, is not restricted to visual sense perceptions. A similar structure characterizes all sensory perceptions, auditory, tactile, etc.[45]. The following quotation from the *Reflections on the Doctrine of a Single Universal Spirit* makes this clear:

---

[44] [128], p. 158

[45] See, e.g., G, IV, 563-564.

> We have an infinity of petite perceptions which we are incapable of distinguishing. A great stupefying roar, as, for example the murmur of a large assemblage, is composed of all the petites murmurs of individual persons which are not noticed at all but of which one must nevertheless have some sensation; otherwise one would not sense the whole. (L, 557; G, VI, 534.)

We can now return to the question concerning the use of a microscope for the purpose of distinguishing details in a phenomenon which were not distinctly perceived. The question is why a microscope can help us see finer details, thus making it possible to distinguish petites perceptions which were insensible to us before its use. It is important to examine, first, what it is to say that a microscope helps us in this way. In view of the above analysis, such a contention amounts to saying that the use of a microscope results in the shifting of the perceptual boundary connected with the representational structure of each one of us. Our visual organs are metaphysically constituted of monads representing in a specific way what we call reality. Our mind, i.e., the dominant monad which is our mind, represents these monads as representing such a reality. There is a neurophysiological process of transmission, i.e., a causally explicable physical process, of the corresponding signal. Such a process is "taking place" in the phenomenal realm and it has its counterpart in the real world of monads representing one another. It is difficult to give an exact model for the correspondence between the phenomenal and the real in this particular case, but for our purposes it is enough to say that such a neurophysiological process can be cashed out in terms of a series of indirect representations which appears to us as temporal.

A characteristic of the complexity of any such series of indirect representations (with what we call external reality as its initial point and our mind as its terminus) is our visual minimum perceptibile. When a microscope is used the whole picture undergoes a dramatic alteration. The monads constitutive of our eyes represent what they were representing before in an unaided way, as represented by the monads constitutive of the microscope. The monads constitutive of the microscope-due, perhaps, to their particular mutual, simultaneous, phenomenal arrangement and to perceptual characteristics of their own-have the ability to represent what was represented before by our unaided eyes in such a way that finer details can be distinguished now by our aided by the microscope eyes. The monads constitutive of our eyes represent the monads constitutive of the microscope as representing external reality distinguishable in finer details. Finally, our dominant monad (i.e., our mind) indirectly represents external reality by also distinguishing finer details. The use of the microscope results in a new finer visual minimum perceptibile for the observer. This is exactly what enables him to distinguish petite perceptions which were insensible before. The finer visual minimum perceptibile is a new boundary, in the sense that the perceiver cannot perceive all of the divisions of the phenomena into lesser ones *ad infinitum*, but he can nevertheless distinctly perceive divisions which could not be distinctly perceived by him before.

Confusedness and distinctness, as characteristics of specific perceptions and more generally of various monadic representational structures, are Leibnizian tools for solving all sorts of different problems in his monadic metaphysics. They are used so extensively by him and with aims of such diversity, that one wonders about their overall effectiveness. They are nevertheless simple and powerful tools, in that they enable him to build quite a novel, impressive, coherent and well-structured metaphysics. It is worthwhile to list two more usages by Leibniz of the confusedness/distinctness dichotomy to indicate the solutions he has to offer to problems that appear in his representational metaphysics.

1. A monad contains in its complete concept, once and for all, all its predicates. Every possible simple substance, as it exists in a primary sense in God's mind, is characterized by such a complete containment. Its predicates,, in a non-temporal sense, constitute, as it were, its blueprint. It is nevertheless the privilege of those monads which have been created by God that they are in particular states as described by such a pre-existing blueprint. Creation of a monad by God is tantamount to its realization in this world which appears as spatio-temporal. The particular details of a monad's unfolding are described by its predicates, but they do not coincide with them. It is exactly this which forced McGuire to say that Leibniz,

> ...can distinguish between the state of a substance at a particular time and the fact that it has an attribute such that it is in that state at that time[46].

It is correct to say that a monad is in a particular state whenever it represents the entire world in its spatio-temporal entirety and its minutest detail in a perspective and confused way. In this context where "perspective" means a way characterized by, among other things, the exact spatio-temporal phenomenal position of the monad as it represents all the others (itself included) as representing all the others. From such a point-of-viewish position each monad represents the entire world, which means that it represents all past and future events as well as every little detail pertaining to the momentary and simultaneous phenomenal arrangement of all other monads (itself included) as if they were positioned in space. In Leibniz's *Clarification of the Difficulties*

46 [103], p. 300.

which Mr. Bayle has found in the New System of the Union of Soul and Body we find the following interesting passage:

> ...the substantial units are nothing but different concentrations of the universe represented according to the different points of view by which they are distinguished. (L, 493; G, IV, 518.)

The point-of-viewish representation of the world is not static. According to Leibniz in every representation, in every point-of-viewish perception of the present, there is a "veritably infinite multitude of little indistinguishable feelings" which are the seeds in the present of what is going to happen in the future. They are indistinguishable little expectations of the *tomorrow*. In the following passage, from the same source Leibniz explains in what way a representation of the present is also a confused and indistinguishable representation of the future:

> I add that the perceptions which are found together in one soul at the same time include a veritably infinite multitude of little indistinguishable feelings, which the subsequent series must develop, so that we need not be astonished at the infinite variety of what must result from it in time. All this is only a consequence of the representative nature of the soul, which must express what happens, and even what will happen to its body and in some way in all other bodies, through the connection or correspondence between all the parts of the world. (L, 496; G, IV, 523.)

In his letter to Johann Bernoulli, dated February 21, 1699, Leibniz returns to the theme of representing the future in the present, making it clear that such a representation is indistinguishable in its details because of the multitude and the smallness of its perceptual parts:

> 'There is also no doubt that our future states are already in some way involved in our present ones, though they cannot be distinguished because of the multitude and smallness of the perceptions occurring at the same time. (L, 5I3; GM, III, 574.)

A similar confused representation of the past in the present is indicated in the following passage from Leibniz's *Reply to the Thoughts on the System of Preestablished Harmony Contained in the Second Edition of Mr. Bayle's Critical Dictionary, Article Rorarius:*

> It is ... these present perceptions, along with their regulated tendency in conformity to what is outside, which forms the musical score which the soul reads. "But", says Mr. Bayle, "must not the soul recognize the sequence of the notes (distinctly) and so actually think of them?" I answer 'No'; it suffices that the soul has included them in its confused thoughts in the same way that it has a thousand things in its memory without thinking of them distinctly.

> Otherwise every entelechy would be God, if it were distinctly conscious of the whole infinite it includes. (L, 580; G, IV, 564.)

Finally, Leibniz insists in the following passage from *The Monadology* that the whole world of monads is represented confusedly in its phenomenal spatio-temporal entirety by each monad:

> The nature of the monad being to represent, nothing can keep it from representing only a part of things, though it is true that its representation is merely confused as to the details of the whole universe and can be distinct for a small part of things only, that is for those which are the nearest or greatest in relation to each individual monad. Otherwise each monad would be a divinity. It is not in the object but in the modification of their knowledge of the object that the monads are limited. ...everybody responds to everything which happens in the universe, so that he who sees all could read in each everything that happens everywhere, and, indeed, even what has happened and what will happen, observing in the present all that is removed from it, whether in space or in time. ...But a soul can read within itself only what it represents distinctly. (L, 648-649; G, VI, 617.)

A monad cannot therefore have a global distinct representation. It can only distinguish finitely many details of things appearing as spatio-temporally near by. In other words, the distinct part of a representation can only be of a finite and local nature. This is not to say that everything in that locality can be represented distinctly. It is, on the other hand, correct to say that what distinguishes the representation of what phenomenally lies beyond the boundary of such a spatio-temporal locality, from the representation of what phenomenally lies inside such a boundary, is their different degrees of confusedness. In the first case, the representation is so confused that it becomes completely unconscious. In the second, the representation, though still confused in its minutest details, is less confused than in the first case, in the sense that finitely many spatio-temporally extended items can be distinguished by the perceiver.

2. Confusedness and distinctness of perceptions are used by Leibniz in one more way, namely, for the purpose of giving an explanatory account of activity and passivity. Following Russell[47], we could say that "in relation to clearness[48] of perception, monads are said to be active or passive," where "this sense of activity must not be confounded with that which is essential to substance," namely, with the essential active principle (appetition) responsible for the

---

[47] [128], p. 141.

[48] A more appropriate word would be "distinctness."

unconscious and spontaneous move of a monad from one representational state to another. According to Leibniz,

> ...anything which occurs in what is strictly a substance must be a case of 'action' in the metaphysically rigorous sense of something which occurs in the substance spontaneously, arising out of its own depths; for no created substance can have an influence upon any other, so that everything comes to a substance from itself (though ultimately from God.) But if we take 'action' to be an endeavor towards perfection, and 'passion' to be the opposite, then genuine substances are active only when their perceptions (for I grant perceptions to all of them) are distinctly better developed and more distinct, just as they are passive only when their perceptions are becoming more confused. Consequently, in substances which are capable of pleasure and pain every action is a move towards pleasure, every passion a move towards pain. (NE, 210; G, V, 195-196.)

It is obvious from this quotation (from the *New Essays on Human Understanding*) that what appears as active is, according to Leibniz, more perfect than what appears as passive. Additionally, activity versus passivity is explicable in terms of distinctness and confusedness of the perceptions of the corresponding active and passive agents. Passivity is a sign of subordination of the corresponding agent to the active one, cashed out in terms of their degrees of perfection. Action and passion appear as causally connected. The metaphysical basis of such a connection is the appropriately coordinated perceptual states of the corresponding agents in a way pre-established by God.

At this point it seems appropriate to ask ourselves what it is that could have led Leibniz to use the dichotomy of distinctness / confusedness of perception as an explanatory metaphysical schema for the connection of action with passion. We have already seen that Leibniz grades monads according to the degree of distinctness of their perceptions, with God's "perceptions" as the absolute, approachable but never reached ideal. Perfection, on the other hand, as an absolute ideal, is an attribute of God. Created beings can be more or less perfect as they are compared to one another, and most certainly less perfect as they are compared to God. Their relative degree of perfection is for Leibniz directly connected with the degree of distinctness of their perceptions. How else could it be, if all there is in a monad is its perceptions and appetition? The more distinct a monad's perceptions the more perfect a monad is. On the other hand, what is active possesses a greater degree of perfection than what is passive. According to Leibniz:

> 49. The created being is said to *act* outwardly in so far as it has perfection, and to *suffer* from another insofar as it is imperfect (L, 647; G, VI, 615.)

The above quoted passage taken from *The Monadology* is quite illuminating. Leibniz tells us in a very clear manner that perfection is connected with action and imperfection with passion. Much earlier in the *Discourse on Metaphysics* Leibniz states that:

> ...when a change occurs by which several substances are affected (as in fact every change affects them all), I believe we may say that the one which thereby immediately passes to a greater degree of perfection or to a more perfect expression, exerts its power, and acts, and that which passes to a less degree makes known its feebleness, and *suffers*. Also I hold that every action of a substance which has perfection implies some *joy*, and every passion some *pain*, and *vice-versa*. (L, 313; R, 267; G, IV, 441.)

But if, on the one hand, relative perfection of created beings is measured by the degree of distinctness of their perceptions and, on the other, is related to action and passion in the above way, then action and passion can be (and are, for Leibniz) explicable in terms of the relative distinctness or confusedness of perceptions of the corresponding agents. It seems, therefore, quite plausible to say that it is through the idea of relative perfection of created beings that Leibniz came to the idea of explaining action and passion the way he did. A monad acts as long as it passes to a greater degree of perfection, and *suffers* as long as it passes to a lesser degree of perfection. Degrees of perfection form a continuous spectrum exactly parallel to the continuous spectrum of degrees of distinctness of perceptions. In a letter of his to Ernst Langrave of Hessen-Rheinfels, dated February 1/11, 1686 Leibniz states:

> The action of one finite substance on another consists only in the increase of the degree of its expression joined to the diminution of that of the other, inasmuch as God has formed them beforehand so that they should agree together. (R, 267, G, II, 13.)

Throughout this chapter we had the opportunity to talk about continuity of the gradation of perceptions or representations according to their distinctness or confusedness. We insisted that according to Leibniz the chain of being is continuous in a way parasitic upon the continuity of such a gradation. We also had the opportunity to refer to the continuity of what is perceived as spatially extended and especially to its main characteristic, namely, actual density. Our main goal, however, was not to analyze the Leibnizian notion of continuity (or to discuss the Leibnizian principle of continuity), but to examine Leibniz's special brand of representational metaphysics. Continuity, on the other hand, as the most important architectonic feature of the Leibnizian metaphysics, deserves to be more thoroughly examined in its own right. There is, of course, an additional reason for doing so. In order to be able to discuss Leibniz's ideas of space and time, the spatio-temporally extended, and its metaphysical correlate, one has to examine both Leibniz's brand of representationalism and his doctrine that nature is continuous in every respect. They are, in a sense, the necessary prerequisites if such a discussion is to get off the ground. The next chapter will therefore be devoted to the examination of various aspects of Leibniz's account of continuity, which, in addition to being interesting in its own right, will be helpful in dealing with the problems of the interconnections between the monadic (i.e., the real), the phenomenal, and the ideal level of the Leibnizian metaphysics of space and time.

CHAPTER II

CONTINUITY

A. THE PRINCIPLE OF CONTINUITY AND ITS RELATION TO OTHER LEIBNIZIAN PRINCIPLES

The principle of continuity is one of the most important architectonic principles of the Leibnizian system. The world, as it is represented by us and as it really is, is continuous. Leibniz explicitly defines this principle or law, as he sometimes calls it, in a number of passages throughout his writings. In the following quotations, the first from the *New Essays on Human Understanding* and the second from a letter of Leibniz to De Volder dated March 24/April 3, 1699, Leibniz states:

> Nothing takes place suddenly, and it is one of my great and best confirmed maxims that *nature never makes leaps:* which I called the Law of Continuity. (NE,56;G, V, 49.)

> *No transition is made through a leap*...this holds, I think, not only of transitions from place to place but also of those from form to form or from state to state. For not only does experience confute all sudden changes, but also I do not think any a *priori* reason can be given against a leap from place to place, which would not militate also against a leap from state to state. (R, 222-223;L, 515-516; G, II, 168.)

The principle or law of continuity was first explicitly defined by Leibniz in 1687 in a paper published in the *Nouvelles de la rèpublique* des *lettres* entitled "Lettre de M.L. sur un principe general utile à l'explication des loin de la nature par la consideration de la sagesse divine, pour servir de replique à la response du R. P. D. Malebranche." Traces of it, however, can be found in his previous writings such as the ones dealing with the infinite divisibility of the spatio-temporally extended, and the continuity of motion (G, IV, 229; G, I, 71-73; F, 180; Cout. OF, 522.). Let us see how the principle was formulated in Leibniz's first effort:

> *When the difference between two instances in a given series or that which is presupposed can be diminished until it becomes smaller than any given quantity whatever the corresponding difference in what is sought or in their results must of necessity also be diminished or become less than any given quantity whatever. Or to put it more commonly, when two instances or data*

approach each other continuously, so that one at last passes into the other, it is necessary for their consequences or results (or the unknowns) to do so also. (L, 351; G, III, 52.)

We can distinguish two different forms that the principle of continuity takes in Leibniz's writings:

(a) Nature never makes leaps[1].

(b) The correlations between related cases in nature which can be considered as values of the same variable x and their consequences are described by continuous functions[2]. The properties of beings, for instance, are continuous functions of their essential determinations. The particularity of any such function depends upon the specifics of the corresponding correlation.

The question that arises at this point concerns the nature of the relation between forms (a) and (b). Let us first determine whether one can derive form (b) from (a). It is obvious and straightforward that if nature never makes leaps, the correlation between cases-which can be considered as values of the same independent variable x-and their consequences have to be described by a continuous function $y=f(x)$. If the range of x were continuous and the range of y were not, then form (a) of the principle would be violated. So the function must be continuous.

This argument shows that the principle of continuity in form (b) is a consequence of the principle in form (a). The converse, however, does not hold. We could have, for instance, a complete set of related cases which by its own nature is discontinuous, connected via a continuous function to a similarly discontinuous set of consequences. The following example from mathematics makes the point clearly.

Let us assume that, given the real line, we define as continuous subsets of it all its possible line segments. Then the subsets $(0,1)\cup(2,4)$ and $(0,1)\cup(4,16)$ are[3] discontinuous. We also consider the function $f:(0,1)\cup(2,4) \rightarrow (0,1)\cup(4,16)$ given by $f(x)=x^2$. This function is continuous, though both the range of f(x) and the domain of x are discontinuous subsets of the real line.

---

[1] This should be understood not only in the temporal sense that at no time does nature make leaps, but also in the more general sense that there are no discontinuities of any sort in nature.

[2] We use the notion of a continuous function, which is a much later development in the history of human thought, in order to capture the Leibnizian idea that in their approach to a limit "consequences or results (or the unknowns)" necessarily behave the same way the "data" do, when they continuously approach their limit. A continuous function $f: A \rightarrow B$, where A and B are, for example, subsets of the real line, is one that for every x_1 that belongs to A and every $\varepsilon > 0$, there exists a $\delta > 0$ such that for every x_2 that belongs to A if $|x_1-x_2| < \delta$ then $|f(x_1)-f(x_2)| < \varepsilon$.

[3] By $(a,b)\cup(c,d)$ we mean all the real numbers between, first, a and b and, second, c and d, with a, b, c, and d not included.

To be exact, the terminology we introduced in order to talk about Leibniz's principle of continuity is not the standard one. We used it and we will continue to use it for reasons of convenience. Yet, some further explanations are needed. We have already used the expressions "if the range of x were continuous", "a complete set of related cases which by its own nature is discontinuous" "we define as continuous subsets", etc. According to modern mathematical terminology spaces, ranges, sets or subsets, which we call *continuous,* should be called *connected* and the ones we call *discontinuous* should be called *disconnected*[4]. The intuitive idea is that such sets or subsets are *all in one piece* or that we can always move from any point of the set or the subset to any other point of the set or the subset through a continuous (uninterrupted) line whose points belong to the set or the subset. In order, of course to, define exactly the notion of a continuous set we have to give it the structure of a topological space. Then, following J. R. Munkres (see [113], p. 147) we can define the notion of a *connected* topological space as follows: "Let X be a topological space. A *separation* of X is a pair U, V of disjoint non-empty open subsets of X whose union is X. The space X is said to be *connected* if there does not exist a separation of X".

Though it is true that (b) is not equivalent to (a), the way we expressed the principle of continuity in form (b) does not do justice to Leibniz. It is an undeniable fact that when he defines the principle he pays some attention to the status of the domain of the independent variable x. We believe he makes the implicit assumption that the domain of the independent variable has to be continuous. Otherwise he would not speak of "two instances or data approach[ing] each other continuously." A stronger evidence, suggesting that he meant that the domain of the variable x is continuous, is the way he expressed a more general principle upon which, he claims, the principle of continuity depends. This more general principle runs as follows: "as the data are ordered, so the unknowns are ordered also."[5] Actually, it is this principle that, taken with a grain of salt, can be said to be the Leibnizian equivalent of what we called form (b) of the principle of continuity. In a sense, it says that correlations between data and unknowns in nature are describable by continuous functions, no matter whether the domain of the data is continuous or not.

In light of this discussion, it seems appropriate that we should modify (b) as follows:

---

[4] See e.g. [89], [113], [134]

[5] "Datis ordinatis etiam quaesita sunt ordinata" (G, III, 52.)

> (c) The correlations between related cases in nature[6]-which can be considered as values of the same variable x-and their consequences, are described by continuous functions. Additionally, the domain of such a variable is always continuous.

It is this form (c) of the principle of continuity that is equivalent to form (a).

We can now turn our attention to the relation of the principle of continuity to some other Leibnizian principles. We will examine, at first, how it is related to the so-called principle of the best. This principle asserts that the created world that we live in is the best possible. The principle of continuity is for Leibniz "a principle of general order" (G III , 52). This actualized world of ours is the best possible and therefore is ordered in the best possible way. The best ordering, Leibniz claims, is the continuous one. So this world has to be governed by the principle of continuity[7].

In the same letter to De Volder mentioned at the very beginning of this section, Leibniz makes an attempt to explain the grounds on which one can uphold this metaphysically and aesthetically important, for the Leibnizian system, principle:

> This is an axiom that I use - *no transition is made through a leap*. I hold that this follows from the law of order and rests upon the same reason by which everyone knows that motion does not occur in a leap; that is, that a body can move from one place to another through intervening positions. I admit that once we have assumed that the Author of things has willed continuity of motion, this itself will exclude the possibility of leaps. But how can we prove that he has willed this, except through experience or by reason of order? ... why could God not have transcreated a body, so to speak, from one place to another distant place, leaving behind a gap either in time or in space; producing a body at A, for example, and then forthwith at B etc.? Experience teaches us that this does not happen, but the principle of order proves it too according to which, the *more we analyze things the more we satisfy our intellect*. This is not true of leaps, for here analysis leads us to mysteries. (L, 515-516; G, II, 168.)

De Volder pressed him for a more satisfactory explanation of the principle as applied to continuous motion. In his reply, Leibniz abandons the appeal to experiential grounds and simply states that,

---

[6] "Nature" is an all-inclusive term referring to all three levels of the Leibnizian metaphysics, the real, the phenomenal, and the ideal.

[7] Discontinuities are irregularities (G, III, 54) and irregularities are signs of disorder and imperfection.

> ...this hypothesis of leaps cannot be refuted except by the principle of order[8], with the aid of the supreme reason, which does everything in the most perfect way. (L, 521; G, II, 193.)

For Leibniz, the principle of continuity is, a beyond doubt, a "principle of general order". Yet this is not enough. What has to be shown,, or reasonably postulated, is that it is the best possible principle of ordering an infinite world. The questions he addresses, in the first quoted passage, are how we know that this world is continuous, and how we can convince ourselves that the best possible ordering of an infinite world is the continuous one. A direct, but not entirely convincing answer to the first question is contained in this same passage. "Experience teaches us" so. That such an answer is not entirely convincing, even to Leibniz, is obvious. In his reply to De Volder he does not return to it and instead asserts that only "the principle of order, with the aid of the supreme reason" can refute "the hypothesis of leaps."

The passage quoted from Leibniz's letter of March 24/April 3, 1699, provides an answer of sorts to the second question, which is also in a sense an answer to the first (since this world is the best possible.) The criterion for determining which is the best possible ordering of an infinite world is *intellectual satisfaction*. Such a satisfaction is experienced through analysis of the continuous. An analysis of leaps, or of the discontinuous, leads to "mysteries," which are sources of dissatisfaction and frustration. Experiencing intellectual satisfaction, Leibniz seems to say, is a sign that we contemplate the best. Although humans are finite perceivers and knowers, they resemble God. In our case we take the intellectual satisfaction of contemplating the continuous as a sign of its being the best among all possible orderings of the infinite. God, on the other hand, loves it because he *knows* it to be the best. If this is so, we can use the criterion of intellectual satisfaction to convince ourselves that the best possible ordering of an infinite world is the continuous one, and that this world is continuous.

Such a psychological criterion has its own familiar shortcomings[9]. In light of the second quoted passage we come to realize that Leibniz drops it and uses instead the "supreme reason" as the support for the principle of continuity. God does everything in "the most perfect way," which means "in the best possible way," with reason. If we were to reconstruct the argument[10] presented

---

[8] "The principle of order" here means the principle of continuity.

[9] Yet, though philosophically weak, it gives us a clue as to how Leibniz came to believe that the best order of the infinite is the continuous.

[10] A rough sketch of which we gave a little while ago.

in this passage by bringing to the fore all of its hidden premises we believe it would take the following form:

God willed, with reason'[11] to create the best of all possible worlds.

The best of all possible worlds is a world ordered in the best possible way.

A possible infinite world which is ordered in the best possible way has to be governed by the principle of continuity.

This world is infinite.[12]

Therefore, this world has to be governed by the principle of continuity.

According to the above argument the principle of continuity is in a fundamental way dependent upon that which we would call ''the principle of the best''. There are a few things which should be noted about such an argument. The first premise is connected with problems concerning God's nature and especially God's free will, which we do not intend to discuss. The second premise has a hidden assumption built into it, according to which possible worlds can be evaluated with respect to the way they are internally ordered. Such an interworld evaluation leads God, or the metaphysician, to single out a possible world ordered in the best possible way. If more than one possible world is ordered in the best possible way, then being the best possible world involves more than being so ordered. But this does not harm premise two. This is so because, to be a world ordered in the best possible way, it is a necessary, but not a sufficient condition, for being the best possible world. The third premise of the argument seems to be the most controversial. It is not metaphysically necessary and it is heavily laden with Leibniz's view of discontinuities as irregularities. Irregularities are for him signs of imperfection and their existence cannot be a characteristic of the best possible world. In other words, interworld evaluation involves already the assumption that the continuous ordering of a world is the most perfect one.

Before we move on to the examination of the relation of the principle of continuity to some other Leibnizian principles, we need to consider premise 4, for this seems to be of special importance in framing the argument we gave. If a world is to be continuous, it has to be infinite. Any finite world will necessarily be non-continuous with respect to its ordering (if there is one.) Continuity involves density, that is, the property according to which between

---

[11] The reason being that it is the best of all possible worlds.

[12] A world can be infinite either with respect to mereology or extension and duration or both. For Leibniz the best of all possible worlds had to be doubly infinite, finitude being for him a limitation and therefore an imperfection.

any two members of a continuous chain there must exist a third member of the chain. In other words, a continuous chain is an infinite chain. This is not to say that an infinite world has to be continuous; it could be discrete with respect to its ordering[13]. But if a world is continuous, it must also be infinite.

A possible world can be infinite without being continuous. The best possible world is infinite (finitude being a kind of limitation and therefore imperfection) and also continuous, because the best possible way to order the infinite is to order it continuously. This is what makes Leibniz say that it [the principle of continuity] "has its origin in the *infinite*" (L, 351; G, III, 52.)

Summarizing our account we can say that for Leibniz the principle of continuity is a consequence of the principle of the best. He seems to think that since the principle of the best is metaphysically necessary, the principle of continuity has to be metaphysically necessary also. But as we saw, its derivation from the principle of the best involves a premise, premise 3, which is not metaphysically necessary, in the sense that discontinuities could be characteristics of the best possible way of ordering an infinitude of things. Premise 3, in other words, is of an architectonic nature and it is a postulate of an aesthetic rather than of a metaphysical origin, in that it pleases the metaphysician.[14]

There are some passages in which Leibniz tries also to connect the principle of continuity to the principle of sufficient reason[15]. The principle of sufficient reason is the second of the two great principles connected with analytic and contingent truths. In the following passage from *The Monadology* Leibniz describes the two principles:

> 31. Our reasonings are based upon two great principles: the first the *Principle* of *Contradiction*, by virtue of which we judge that false which involves a contradiction, and that true which is opposed or contradictory to the false;
> 32. and the second the *Principle* of *Sufficient Reason*, by virtue of which we observe that there can be found no fact that is true or existent, or any true proposition, without there being a sufficient reason for its being so and not otherwise, although we cannot know these reasons in most cases. (L, 646; G, VI, 612.)

---

[13] The Leibnizian real world of monads is infinite. A superficial examination of its structure could lead one to the impression that it has the characteristics of the structure of a discrete and discontinuous world. As we shall see in chapter III, such a view is completely mistaken.

[14] For a similar view, see [128], p. 64.

[15] As, for instance, when he insists that discontinuities in nature "would upset the great Principle of Sufficient Reason and would compel us to have recourse to miracles or to pure chance in the explanation of phenomena" (BC, II, 557.)

In his correspondence with Clarke (Leibniz's "Second Paper") he insists:

> The great foundation of mathematics is the Principle of Contradiction, or Identity, that is, that a proposition cannot be true and false at the same time; and that therefore A is A and cannot be not-A ... But in order to proceed from mathematics to natural philosophy another principle is required ... I mean the Principle of Sufficient Reason, viz., that nothing happens without a reason why it should happen and not otherwise. (A, 15-16; G, VII, 355-356.)

The argument that Leibniz uses in order to derive the principle of continuity from the principle of sufficient reason is simply that God would not have a sufficient reason for choosing to actualize a discontinuous world over a continuous one. As it stands the argument is brief and quite weak. There are some hidden premises which we have to consider before we can properly evaluate it.

It is generally accepted by Leibniz scholars that more than one version of the principle of sufficient reason can be distinguished in Leibniz's writings[16]. The principle in one of its versions states that it is final causes which provide the sufficient reason for things as they really are or as they are faithfully represented. A final cause for the actualization of this world is God's perfectly reasonable decision to create it. God's reason for creating it was that it is the best possible, and such a reason is sufficient. In other words, God would not have a sufficient reason for choosing to actualize a world which is not the best possible. As Leibniz himself puts it (*The Leibniz-Clarke Correspondence,* Leibniz's "Fifth Paper"):

> ...a contingent which exists, owes its existence to the principle of what is best, which is a sufficient reason for the existence of things. (A, 57; G, VII, 390.)

We are now in a position to reconstruct Leibniz's argument as a *reductio ad absurdum:*

> This world of ours is not governed by the principle of continuity.
>
> An infinite world which is not governed by the principle of continuity is not ordered in the best possible way.
>
> A world which is either finite or not ordered in the best possible way is not the best possible.
>
> God willed to create this world.
>
> The only sufficient reason for the actualization of a world is that it is the best possible.

[16] See, e.g., [128], pp. 30-35, and [106], pp. 104-110.

Therefore, God acted without sufficient reason by actualizing this world.

This argument has shortcomings similar to those we discussed in the case of the argument concerning the derivation of the principle of continuity from the principle of the best. This was to be expected, since what we really did was to consider additionally the interconnection between the principle of sufficient reason and the principle of the best, in light of the passage quoted above (A 57; G, VII, 390.)

Finally, we will examine the relation of the principle of continuity to two more principles which seem to be immediately connected with it, the principle of the identity of indiscernibles[17] and the principle of plenitude. The principle of the identity of indiscernibles-or of the dissimilarity of the diverse, as McTaggart called it-asserts, in the *Discourse on Metaphysics*, that "it is not true that two substances resemble each other completely, and differ only numerically" (L, 308, G, N, 433.) More precisely, there are no two different monads with exactly the same complete notions. The principle of the identity of indiscernibles applies to substances only and not to existent attributes. Paired with the principle of continuity, it implies that this world is not only a plenum, but a plenum containing the greatest possible number of coexisting substances, which are unique in the sense that no replica of any one of them coexists with them.

The principle of the identity of indiscernibles and the principle of continuity are mutually independent. We can imagine a world consisting of finitely many individual, undivided and partless substances differing from one another via the containment of different predicates which also could be finitely many. Such a world can be imagined as being analogous to the simple model of a finite group of differently colored marbles. It would then be a world in which the principle of continuity is violated and the principle of the identity of

---

[17] According to Russell we can extract from the Leibnizian writings the following argument in support of this principle:

> All that can be validly said about a substance consists in assigning its predicates. Every extrinsic denomination-i.e., every relation-has an intrinsic foundation, i.e., a corresponding predicate (G. II. 240). The substance is, therefore, wholly defined when all its predicates are enumerated, so that no way remains in which the substance could fail to be unique. For suppose A and B were two indiscernible substances. Then A would differ from B exactly as B would differ from A. They would, as Leibniz once remarks regarding atoms, be different though without a difference (N.E. : p. 309; G. V. 268). Or we may put the argument thus: A differs from B, in the sense that they are different substances; but to be thus different is to have a relation to B. This relation must have a corresponding predicate of A. But since B does not differ from itself, B cannot have the same predicates. Hence A and B will differ as to predicates, contrary to the hypothesis. ([128], p. 58.)

indiscernibles holds[18]. Conversely, we could imagine a continuous world containing non-identical indiscernibles. For instance, we can imagine a world differing from the actual[19] in exactly one detail. For a particular monad A there exists an exact duplicate a monad B in the following sense: monad A by representing itself represents monad B and monad B by representing monad A represents itself. Then monads A and B have indiscernible representational structures, but they are not identical[20]. All the other monads represent the world as if monads A and B were identical[21]. In the world we just described the principle of the identity of indiscernibles is violated and the principle of continuity holds.

According to the above analysis, the two principles are mutually independent. Yet they are closely connected, because, as Russell puts it, both are

> ... included in the statement that all created substances form a series, in which every possible position intermediate between the first and last terms is filled once and only once. That every possible position is filled once is the Law of Continuity; that it is filled only once is added by the Identity of Indiscernibles. ([128], p. 54.)

Let us now turn our attention to the principle of plenitude. This is the principle according to which the actual world is inhabited by the largest possible number of actualized substances, so that there is no vacuum either at the metaphysical (i.e., the real) or the phenomenal level. "God does everything," says Leibniz, "in the most perfect manner possible, and a *vacuum* of *forms exists* no *more than* a *vacuum* of *bodies*" (M, 160; G, II, 125.) At the metaphysical level substances form a complete continuum. At the level of the

---

18 The principle of continuity would be violated because the existing substances would not be able to form a continuum.

19 Which is continuous, according to Leibniz. If this is not entirely satisfactory we can imagine in its place a continuous world with no indiscernibles.

20 Such an example raises profound metaphysical problems, as to whether there could be a possible world in the Leibnizian pantheon of possibilia in which substances differ *solo numero*, which we do not intend to discuss here. It is enough for our purposes to observe that there is a theoretical possibility of such a world.

21 In other words, although monads A and B have exactly the same complete notions and they are represented by all the others as identical, they are still non-identical inhabitants of the same world. The important thing is that monad A does not represent monad B as differing in any way from A itself and vice-versa. So the premises of the argument that Russell attributes to Leibniz in the case of the identity of indiscernibles (footnote 17) do not hold here. In other words, that A and B are not identical is not cashed out in terms of a predicate contained in A and a predicate contained in B, which by spelling out the difference between A and B would also make A and B different from each other. That A and B are indiscernible, but non-identical is something that only God could know and the metaphysician could assume and postulate.

phenomena there are no vacua or hiatuses. For instance, to every spatial point at a given time t[22] corresponds a monad representing the world from its point of view; it is as if the monad were positioned without being really situated at this spatial point viewing the world perspectively. Conversely, to every monad there corresponds at a given time t a unique point of view which can be said to be, among other things, the representational and perspective analogue of a spatial point. So what appears as a spatial configuration of simultaneously coexisting entities is a plenum, since spatial vacua would correspond to a set of points of view of non-existing monadic representors. A spatial vacuum, in other words, would correspond to a vacuum of substances and therefore would lead to the idea of an existing world less perfect than it should be.

The question we have to face at this point is how the principles of plenitude and continuity are connected. Lovejoy believed the latter to be a mere consequence of the former[23]. We think he was mistaken. The principle of continuity is not a consequence of the principle of plenitude. This is true even under the assumption that we can consider only infinite worlds as our objects of investigation. As we shall show, the converse implication obtains.

Let us consider an infinite possible world[24] which is discrete (and therefore noncontinuous) in the following sense: it consists of substantial atoms which occupy or represent and are represented as occupying every position in a discretely structured simultaneous mutual arrangement. Such an arrangement could resemble a three-dimensional infinite matrix, which is isomorphic to the Cartesian product Z^3, where Z is the set of integers[25]. Every substantial atom occupies or represents and is represented as occupying an absolute position characterized by an ordered triple (a, b, c), where a, b, c are integers[26]. The relative position of such an atom is characterized by the triple (a, b, c), specifying its absolute position, together with the total mutual configuration of all the other atoms as they occupy or represent and are represented as

---

22 In chapter III we will see in more detail what a spatial point and an instant of time are, and which metaphysical entities correspond to them.

23 See[95], p. 145, note 2.

24 This is an arbitrary construct and should not be interpreted as referring to Leibniz's idea of the best possible world we are living in. In the past a rather successful attempt has been made towards constructing or giving a mathematical interpretation of Leibniz's universe (see [60]). Our construction was inspired by it.

25 Given a non-empty set A the Cartesian product A^n, with n a natural number greater than or equal to one, is the set of all ordered n-tuples with entries from A.

26 We use the phrase "represents and is represented as occupying" in order to include the possibility of a world where positioning, moving, etc., are phenomenal as in the case of the Leibnizian world of the usual spatio - temporal phenomena.

occupying all the other absolute positions at a particular state of the world. A state of the world (real or phenomenal) should be thought of as characterized only by the simultaneous mutual arrangement of the substantial atoms corresponding to it. Change (real or phenomenal) is also discretely structured, in the sense that such a simultaneous mutual arrangement is followed always by some next one[27], without any intermediate simultaneous mutual arrangements occurring between them. Additionally, change from a state of the world to its next one means that each atom during such change either stays still or moves from its position to a position absolutely next to it. As an abstract notion, change would be represented isomorphically as an exhaustive enumeration of the integers. In other words, we could imagine the total history (real or phenomenal) of such a world as isomorphic to a linear discrete ordering of the set of all simultaneous mutual (real or phenomenal) arrangements of the substantial atoms, which would occur during the enfolding of such a history. As a consequence, we would have a complete one to one correspondence between the totality of substantial units and the set of all personal histories, where a personal history is a discrete linear ordering, isomorphic to Z, of the set of relative (real or phenomenal) positions of a substantial atom in the framework of the total history of the world.

If change could, as suggested, be thought of as isomorphic in its flow to an exhaustive enumeration of the integers, then to every state (real or phenomenal) of the world would correspond an integer and, to every integer, a state of the world. Time intervals between any two states of the world would be measured by the positive difference $|m-n|$ where m and n are the integers corresponding to these states. Spatial distance between any two substantial atoms would be measured by the positive integer $d = |c-g| + |b-f| + |a-e|$, where (a, b, c) and (e, f, g) are the absolute positions of the corresponding substantial atoms. Motion could then be conceived of as the result of discrete changes in the simultaneous (real or phenomenal) mutual arrangement of the substantial atoms. Finally, we could postulate that such a world could be at a particular state only once during its total history. We should add at this point something which is of a technical nature. Such an addition is necessary for without it the above described construction would probably appear incomprehensible to the reader. A construction like the above is possible because the cardinality of the

[27] Every simultaneous mutual arrangement of the substantial atoms satisfies the condition that for every absolute position, as described by the triple (a, b, c), where a, b, and c are integers, there exists a substantial atom occupying it really or phenomenally and that every substantial atom occupies really or phenomenally such a position.

sets constructed is that of the integers. We also used the integers in order to express the idea that the history of such a world has no beginning and no end.

There is no need to make the picture of such a world clearer. The important thing is that the possible world we described is a plenum. This is so because no vacuum exists with respect to substantial atoms, or with respect to what we would call, in the case of such a world, space and time. Substantial atoms form a plenum because, within the framework of the total history of the world, for every possible personal history there is a unique substantial atom with such a history and vice-versa. A vacuum of substances would correspond, in other words, to a set of personal histories with no substantial atoms having them, which is not the case for the world we described. Spatio-temporally, the world is a plenum, because a vacuum in space or time would exist only if spatial or temporal positions were not occupied by substantial atoms, or did not correspond to different states of the world. More specifically, the existence of a vacuum in time would be tantamount to the existence of a time interval not equal to zero with no change corresponding to it. The existence of a vacuum in space, at a particular state or states of the world, would be equivalent to the existence of spatial positions not occupied by corresponding substantial atoms. Additionally, any substantial atoms when situating next to one another are not separated by a spatial vacuum, because there are no intermediate (real or phenomenal) positions between the ones that these atoms occupy. Similarly, any two states of the world that are next to one another are not separated by a temporal vacuum, because time is not running independently of change, but is parasitic upon it, since it is defined as the measure of change.

The world we described is a plenum. On the other hand, it is non-continuous, because it does not satisfy a condition which is a basic characteristic of continuity. A continuous world is a dense world, i.e., it is a world characterized by the existence of intermediaries between any two different elements of a chain (substantial or other) in the world.

Our initial aim was to show that the principle of continuity is not a consequence of the principle of plenitude. The theoretical possibility of a world like the one we described is not to be excluded. Additionally, the construction of such a possible world can be thought of as a construction of a counterexample to the claim that the principle of continuity can be derived from the principle of plenitude.

That the principle of plenitude is a consequence of the principle of continuity is easy to show. If the principle of plenitude did not hold, then the world would contain vacua. But vacua are gaps in the order of the world, and such a thing would contradict the maxim that "nature never makes leaps." In short, the right way to see the connection between the two principles is through the irreversible sequence: *continuity, therefore no gaps, and consequently fullness of the world.*

B. KINDS OF CONTINUITY AND APPARENT CASES OF DISCONTINUITY

In order to show the power and importance of the principle of continuity in Leibniz's system, we will present two examples of how he used it as *a priori* ground for, first, refuting the Cartesian rules of motion and, second, for proving the non-existence of material, extended atoms. The first two Cartesian rules of motion, phrased in the idiom of today's elementary mechanics of absolutely elastic collisions, can be stated as follows:

> *Rule* 1: If two bodies B and C of equal mass, moving with equal speeds in opposite directions, collide directly, then both will be deflected and move with the same speed in directions opposing the ones they were following before the collision.
>
> *Rule* 2: If two bodies B and C of unequal masses ($m_B > m_C$), moving with equal speed in opposite directions collide directly, then they will both move with the same speed as before in the direction that the body with the greater mass was moving before the collision.

Leibniz agrees that rule 1 is correct, as it indeed is. But he goes on to prove that, given the correctness of rule 1, the principle of continuity forces us to accept that rule 2 is false. Considering a case described by rule 2, he argues convincingly, in a passage from the *Letter of Mr. Leibniz on a General Principle Useful in Explaining the Laws of Nature Through a Consideration of the Divine Wisdom; to Serve as a Reply to the Response of the Rev. Father Malebranche*, that:

> ... the inequality of the two bodies can be made as small as you wish and the difference between the assumptions in the two cases, that is, the difference between such inequality and perfect equality, becomes less than any given difference. Yet, if Rule 2 were true as well as Rule 1, the result would be the contrary, for according to Rule 2, any increase, however small. of the body B, formerly equal to C will make the greatest difference in the effects, in that it will change an absolute regression into an absolute continuation of motion. (L, 352; G, III, 53.)

Leibniz continued his criticism of the remaining Cartesian rules of motion along the same lines, with great effect. The second example concerns Leibniz's proof that there are no material extended atoms[28]. The proof presupposes that an atom is absolutely rigid. The reason is that Leibniz connects the non-rigidity of an extended material object with the existence of parts. He seems to think that any change of the shape of an object could only result from a spatial rearrangement of such parts. Non-rigidity, in other words, implies, in Leibniz's eyes, the existence of parts; therefore, material, extended atoms can only be thought of as having an unchangeable shape.

Let us consider two material, extended atoms A and B, moving with the same speed in opposite directions, colliding with each other. Then, if they were completely inflexible the change of their velocities during the moment of their collision would obviously occur through a leap. This, of course, is not permitted by the principle of continuity. What happens, says Leibniz, is that from the moment they touch, the two atoms, start getting gradually and continuously compressed, like two inflated balls. Their centers approach each other more and more closely, until their velocities get the value zero. After that, they rebound from each other in a retrograde motion starting from rest. They increase their velocity continuously, finally regain their normal shape, and move away from each other in opposite directions. The conclusion drawn by Leibniz is that since any collision involves a change of shape of the colliding material extended objects, material extended atoms do not exist.

The principle of continuity, as a "principle of general order," is applicable to every level of the Leibnizian metaphysics of the best of all possible worlds. It is an *organizational*, architectonic principle. It applies equally to the real and to the phenomenal, and even to the ideal. It is at the level of the ideal that the principle is applied when Leibniz insists that:

> ... equality can be considered as an infinitely small inequality, and inequality can be made to approach equality as closely as we wish. (L, 352; G, III, 53.)

or that:

> ... a given ellipse approaches a parabola as much as is wished, so that the difference between ellipse and parabola becomes less than any given difference ... And, as a result, all the geometric theorems which are proved for the ellipse in general can be applied to the parabola by considering it as an ellipse one of whose foci is infinitely removed from the other, or (to avoid the term "infinite") as a figure which differs from some ellipse by less than any given difference. (L, 352; G, III, 52.)

The world is continuous in all its facets. Differences in gradations of related cases in this world can be as small as we please, in the sense that, for every

---

[28] See part II of *Specimen Dynamicum* (GM, VI, 246-254.)

difference between any two elements A and B of a linearly ordered chain of such cases, there exists a third element C, such that its difference from either A or B is strictly less than the difference between A and B.

According to Russell[29], we can distinguish three kinds of continuity in Leibniz's writings: (1) spatio-temporal continuity, (2) continuity of cases, and (3) continuity of actual existents or forms. We think the division is not exhaustive and that we can distinguish at least one other kind of continuity, which might be called (4) continuity of existence *ad infinitum*. Spatio-temporal continuity, as Russell observes, is indeed two-fold. It concerns, first, the continuity of what appears as existing in space and time, and, second, the continuity of space and time themselves. Spatially extended objects belong to the realm of the phenomena which are *bene fundata*, because they correspond to the reality of a world inhabited by monadic representors representing one another in a consistent way[30]. An extended object is continuous in the sense that subdivisions can be carried out anywhere throughout its body *ad infinitum*. In a way, the metaphysical basis for carrying out such divisions is already present. Subdividing a spatially extended object does not mean we actualize a potentiality of subdivisions in the Aristotelian sense. It simply means that we make possible the distinct observation of two separate parts of the same object, which were there before the subdivision, but which could not be observed as separate because of their being infinitesimally close to one another. By the expression "infinitesimally close" we mean that between these two separate parts there was no vacuum. This is analogous to the way we conceive the subdivision of a straight line segment into two other straight line segments. In this case, points are there before the act of separation or division, in the sense that from the division two line segments result only one of which has as one of its end-points the pre-existing point in the initial line segment that occupied the position where the division or the cut took place.

Duration of a spatially extended object or of a group of spatially extended objects is also infinitely divisible. Once more, we should not understand subdivisions of duration in the Aristotelian sense; rather, duration consists in distinctly representing change as divided into parts already separate and

---

[29] [128]. pp. 63-65

[30] In chapter I we had the opportunity to examine continuity of the spatially extended, in a preliminary manner, as related to confused perceptions. In chapter III we will return to the discussion of the spatio-temporally continuously extended and will give a more thorough account of what the connection is between it and its metaphysical (i.e., real) correlate, as based upon representation.

infinitesimally close to each other. To complete the picture,[31] we should add that continuity (and therefore infinite divisibility) of what appears as existing in space and time, also governs motion, as well as any other sort of physical process. The continuity of cases is best exemplified in Leibniz's proof of the invalidity of rule 2 of the Cartesian rules of motion and in the proof of the non-existence of material extended atoms, which we already discussed. Other examples of applying what Russell calls continuity of cases are (i) a circle or a parabola as limit cases of a continuous gradation of ellipses, (ii) rest as a limit case of motion, (iii) equality as a limit case of inequality, etc. In all of these examples the important thing is that the properties of the limit case are the appropriate limits of the properties of the cases which approach the limit case. Continuity of cases, in other words, is a clear manifestation of a basic characteristic of interconnections between cases which form ordered series and their results. Such interconnections are describable by continuous functions. If we were to phrase loosely the situation Leibniz describes, we could say that *nature respects limits*. According to Russell, continuity of cases is the "sole form" of what we called the principle of continuity in form (b):

This principle states, that when the difference of two cases diminishes without limit, the difference in their results also diminishes without limit, or, more generally, when the data form an ordered series, their respective results also form an ordered series, and infinitesimal differences in the one lead to infinitesimal differences in the other. ([128], p. 64.)

We can now turn our attention to the continuity of actual existents or forms. As we have argued in chapter I, beings form a continuous spectrum. Such a spectrum runs parallel to the continuous spectrum of degrees of confusedness or distinctness of the representational structures of those beings, that is, of the representational structures of their dominant monads. From our point of view, Leibniz was mistaken in thinking that there is a unique chain of beings and that such a chain is continuous. But this is not important. Our primary aim is not to prove that Leibniz was trying to explain too much too soon, but to see what he was committed to when he put forward the maxim that "nature never makes leaps."

We should, finally, consider what we called continuity of existence *ad infinitum*. According to Leibniz the fundamental substantial units (i.e., the monads) are everlasting. They are not generated at a particular time out of nothing, and similarly they do not cease to exist. Both generation and

---

[31] About the continuity of space and time themselves we will speak in chapter III. For the moment it suffices to say that in this case Leibniz gives us an account of continuity quite similar to Aristotle's.

corruption-in the sense of ceasing to be-of fundamental substantial units constitute violations of the principle of continuity, because both amount to the existence of inexplicable leaps in the order of the best possible world.

It is interesting to observe that the principle of continuity plays here a role quite different from the one it plays in the three kinds of continuity already considered. Density or the respecting of limits does not seem to be that important in the case of continuity of existence *ad infinitum*. What is important can be traced to a requirement for an uninterrupted, beginningless monadic life span. Yet, the requirement for an uninterrupted, beginningless, and endless monadic life span, though a consequence of the principle of continuity, could also be fulfilled even in a discretely ordered world as it has already been shown in the previous section where such a world was described. There is a sense, of course, according to which generation and ceasing to be can be seen as violations of density or the respecting of limits. We could, for example, consider existence and non-existence as two absolutely adjacent points in a chain of changes, in which case moving from existence to non-existence, or vice-versa, would violate the requirement for density.

Leibniz's theory of transformation of bodies and his denial of transmigration of souls is directly related to the principle of continuity. It is, in a sense, a by-product of the continuity of existence *ad infinitum* and involves a slightly stronger requirement, namely, that what is, is everlasting and keeps intact its basic organizational structure. According to Leibniz, every animal has a soul (i.e., its dominant monad) and a body. The body is in continuous change, but the organism remains the same. The body can diminish or grow, or undergo some other change (observable or not), but the organizational structure of the animal remains intact forever. The soul never leaves the body, it always remains in the same animal, i.e., it remains always in the *same* organizational structure, which can be cashed out in terms of a particular representational structure of the soul. According to Leibniz (letter of his to Nicolas Remond, dated February 11, 1715) the principle of continuity does not leave room for metempsychosis, because:

> ... it demands that everything should be explicable distinctly and that nothing should take place in a leap. But the passage of the soul from one body to another would be a strange and inexplicable leap. What happens in an animal at present happens in it always; that is, the body is in continuous change like a river, and what we call generation or death is only a greater or quicker change than ordinary, as would be a waterfall or cataract in a river. (L, 658; G, III, 635.)

So there is no transmigration of souls, but only transformation of bodies in such a way that the soul, during its uninterrupted, beginningless, and endless life, remains in the same animal. This is what the principle of continuity dictates.

Though Leibniz claims that this world is continuous, there are cases in which appearances deceive us into thinking otherwise. There is a sense in which the realm of phenomena appear to us as discontinuous. Distinct perceptions of extended objects presuppose more or less sharp demarcations between what is perceived distinctly and what is perceived confusedly or unconsciously. Such demarcations give the impression of leaps, which are not in fact to be found in nature, either as it really is or as it is represented. Nature is a plenum and, as such, it does not contain vacua. When we represent the real world of monads representing one another we do it faithfully. A faithful representation is one that does not contain sudden, inexplicable leaps. The metaphysical -i.e., the real- and the phenomenal, are in one sense isomorphic, and continuity of the one presupposes continuity of the other[32]. The impression of discontinuity that we get by perceiving distinctly only some of our perceptions, should not deceive us. The appearance of sharp demarcations, and therefore of leaps in phenomena, is necessary for making the right distinctions and classifications out of the massive, undifferentiated, and continuous web of our representations. Such an appearance, on the other hand, is explicable in terms of the Leibnizian machinery of distinct, confused, and unconscious perceptions. This ability of ours to demarcate phenomenal reality by having distinct, confused and unconscious perceptions, is responsible for both our aesthetic appreciation of what we perceive as reality and our false impression that nature is discontinuous. A passage from the *New Essays* makes this point clearly:

> In nature everything goes by degrees and nothing by leaps, and this rule as regards change is part of my law of continuity. But the beauty of nature, which desires distinguished perceptions, demands the appearance of leaps and of musical cadences (so to speak) amongst phenomena and takes pleasure in mingling species. Thus, although in some other world there may be species intermediate between man and beast...and although in all likelihood there are rational animals somewhere which surpass us, nature has seen fit to keep those at a distance from us so that there will be no challenge to our superiority on our own globe. (NE, 473; G, V, 455.)

Sharp demarcations in the framework of one and the same representation of what really exists should not fool us. We are faithfully representing what really

[32] In chapter III we will see how such an isomorphism can be understood.

exists, but characteristics of the representations, such as distinctness and confusedness (or even unconsciousness) are responsible for the appearance of gaps. What is, is not only a plenum but is also represented as a plenum. That we are not aware of it as such is due to the demarcations in one and the same representation for which the distinctness / confusedness contrast is responsible. Furthermore, the requirement of a continuous gradation of representations according to their distinctness or confusedness does not entail that all possible gradations should be found in one and the same representation of the world. Some could be missing (to be found in some other representation or representations), making the contrast between the distinct and the confused (or even the unconscious) more vivid.

According to the passage quoted above, another source for the appearance of gaps in nature is that we have not observed (or we cannot observe) all possible gradations of beings. This is not so because such beings do not or cannot exist, but because they have been kept "at a distance from us" by nature, for particular reasons. The chain of beings, as they are represented, is continuous, but, although the whole chain is represented in one and the same representing, only finitely many members of the chain can be distinctly perceived. What Leibniz wants to say is that beings, in this actualized possible world of ours, form a continuous chain, whether or not we experience it as such, by having a distinct representation of every one of its members, or of it as a whole. That there are beings which have been kept "at a distance from us" can be explained metaphysically by an appeal to characteristics of the representational structures of us as representors. It is therefore the case that the appearance of gaps in the order of beings is a false impression, and does not constitute a violation of the principle of continuity.

C. DENSITY AND SEQUENTIAL OR CAUCHY COMPLETENESS

The objective in this section will be to see what the basic characteristics of continuity are, from a modern point of view, and whether we can find traces of them in Leibniz's philosophical writings. It will be argued that not only density, but also what we would today call sequential or Cauchy completeness, is in one sense present in Leibniz's treatment of the continuous. We will use what mathematicians call the "real line" as the paradigm for our investigation.

There are two alternative routes for describing the real line. The first is to see it as an infinite set of unextended points — i.e., of entities with measure equal to zero — which is organized through a set of properties and relations of these points into a linearly ordered, extended, continuous whole. The second is to see it as a set of already linearly extended parts -its line segments — overlapping or not, organized, also, through a set of properties and relations of these parts, into a linearly ordered, extended, continuous whole. In the first case our unanalyzable primitives are all the unextended points which constitute the line, and in the second, all its extended line segments. If we adopt the first alternative we can define line segments as sets of points satisfying certain properties. If we adopt the second we can define points as equivalence classes of appropriately "nested" sets of line segments. Both alternatives can be given an appropriate rigorous formalization. These formalizations are equivalent, in the sense that they are intertranslatable, so that any proof of a sentence A in the first formalization can be translated into a proof in the second formalization of a sentence B, which is a translation of A into the framework of the second formalization; and similarly for any sentence provable in the second framework. Let us note here that although these alternative routes are technically equivalent, they are based on different ontological commitments, which the average mathematician is unaware of or indifferent to[33]. Since the first formalization is more popular and since whatever we can say about continuity using its idiom can be translated into the idiom of the second formalization, and vice-versa, we will adopt it without any further comments concerning their interconnection.

The two basic characteristics of the real line, which make it a linearly ordered continuum, are *density* and what we call *sequential* or *Cauchy completeness*. Density is the property according to which for every two points (or real numbers) a and b there is a point c, different from a and b, situated between them; i.e., there is a point c such that $a<c$ and $c<b$. Sequential or Cauchy completeness[34] can be formalized via the *Cauchy-Bolzano-Weierstrass* notion of *convergence* of *sequences*. Let us consider a sequence of real numbers $a_1, a_2, \ldots, a_n, \ldots$, and let us assume that the difference between its elements a_m and a_k approaches 0 as the indices m and k become bigger and bigger. Then, according to sequential or Cauchy completeness of the real line, there is a real number a such that the difference $a_m - a$ approaches 0 as m becomes bigger and bigger. More precisely, a sequence $a_1, a_2, \ldots, a_n, \ldots$, of real numbers is called a *Cauchy sequence if:*

---

33 The first route is connected to a point-set ontology, the second to a line segment set ontology, which is nearly Aristotelian, in the sense that points are parasitic upon the partition of an extended whole prior to them.

34 An alternative and equivalent way for axiomatizing the real line is via the so-called *Dedekind cuts*. We have chosen sequential or Cauchy completeness because we consider it intuitively closer to Leibniz's non-rigorous, fragmentary and not fully realized approach to the idea of what we would today call "genuine" continuity.

$$\forall\varepsilon\exists n\forall m\forall k(m>n\wedge k>n \rightarrow |a_m - a_k| < \varepsilon),$$

where ε is a positive real number and m, n, k are natural numbers. A sequence $a_1, a_2, ..., a_n, ...$, of real numbers is a *convergent* sequence, if there is a real number a, called its *limit* of *convergence*, such that:

$$\forall\varepsilon\exists n\forall m(m>n \rightarrow |a_m - a| < \varepsilon).$$

where ε is a positive real number and m, n natural numbers. It can be proved that every convergent sequence has a *unique* limit of convergence. The property of sequential or Cauchy completeness is expressed by an axiom (written in a second-order predicate language), according to which every Cauchy sequence of real numbers is convergent. What the property entails is that the real line contains no gaps with respect to limits of Cauchy sequences, or, to put it more succinctly, the real line is closed under limit operations. In other words, it entails that whenever accumulations of points exist (and this is the case everywhere in the real line), there is a real number to serve as the point around which the points involved in the particular accumulation are crowded together.

Both density and sequential or Cauchy completeness are properties of all the traditional, paradigmatic continua, such as the real line, the plane, or the three-dimensional Euclidean representation of space. Yet, considered in general, they are independent of one another. That the property of sequential or Cauchy completeness is not a consequence of density can be seen quite easily. It is enough to show that there is a mathematical structure, wherein density holds and sequential or Cauchy completeness fails. The real line contains proper subsets which, considered by themselves, constitute such structures. The most familiar example is that of the set of rational numbers. A *rational* is any real number which can be expressed as m/n, where $n\neq0$, and m, n are integers with no common divisors except 1. The set of rational numbers is dense (i.e., for any two rationals p and q there is a rational r, such that $p<r$ and $r<q$) without being sequentially or Cauchy complete. For instance, we can construct a Cauchy sequence of rational numbers having as its limit the irrational, algebraic, real number $\sqrt{2}$. Since $\sqrt{2}$ is not a rational number, we conclude that the property of sequential or Cauchy completeness is violated, because there is at least one accumulation of rational numbers with no rational number to serve as its focus (i.e., its limit). A similar example can be given if we consider the set of algebraic numbers or the set of transcendental numbers

(which, by the way, is much bigger than the set of rational numbers) by themselves. An *algebraic* number is a real number which is a root of a polynomial equation

$$a_n x^n + a_{n-1} x^{n-1} + \ldots + a_1 x + a_0 = 0,$$

where $a_0, \ldots, a_n$ are integers and n a natural number. A *transcendental* number is a real number which is not algebraic, as, e.g., π and e.

It is also true that density is not a consequence of sequential or Cauchy completeness. In order to show this, we will use the same technique as above, namely, that of giving an example of a mathematical structure wherein sequential or Cauchy completeness holds and density fails. Let us consider the subset of the real line which is described as the set of reals less than or equal to 0 together with the set of reals greater than or equal to 1. Sequential or Cauchy completeness holds, because any Cauchy sequence with elements from the set converges to an element in the set. Density fails simply because between 0 and 1 there is no element c of the set such that $0<c$ and $c<1$.

Continuity for Leibniz is a general architectonic feature of the best of all possible worlds. Both density and sequential or Cauchy completeness are characteristics of the Leibnizian continua, given that these continua are the traditional, paradigmatic ones, as, for example, are the real line, the plane, or the three-dimensional Euclidean representation of space. Yet it is correct to say that for Leibniz continuity meant, most basically, density[35]. We can distinguish in his writings at least two types of density. The first refers to both the real and the phenomenal levels of his system and the second to the ideal.

1. A spatially extended object is continuous, in the sense that it is infinitely divisible. But the metaphysical basis for such divisions does not consist in a mere potentiality. The spatially extended is not potentially divisible in the Aristotelian sense. *It is actually divided.* A spatially extended object is not real. It belongs to the realm of phenomena. That is to say, a spatially extended object is a representation-confused in its details, but distinct as a whole-of an infinite, discrete multitude of monads, which are represented as if they were positioned, relative to one another, in what we would call, ideally, spatial points. The important characteristic of such a representation is that the discrete infinite multitude of monads we are talking about is represented as if the

---

[35] Consider, e.g., how he defines *continuous* extension:

> There is continuous extension whenever points are assumed to be so situated that there are no two between which there is not an intermediate point (R, 247; G, II, 515.)

monads were densely arranged. This means that whenever we represent monads A and B we always (confusedly or unconsciously) represent an infinitude of other monads as if they were positioned in between[36] them. So density is, in a very specific sense, an *actual* feature of the phenomenally extended; it is a well-founded appearance.

Similar points hold for all the Leibnizian continua which belong to the realm of well-founded phenomena. Take, for example, the so-called continuum of beings. According to Leibniz, beings form a continuum which is *actually* dense, because in between any two different elements of this continuum there exists a third, which is less perfect than the first one and more perfect than the second. Beings, as represented, form a continuous spectrum of organic wholes (plants, animals, etc.) following the continuous spectrum of the gradation of their dominant monads.

2. Space and time as ideal notions or, for that matter the real line, the plane, and the space of the three-dimensional Euclidean geometry are continua of a different sort. The real line, for example, is a mental entity prior to its parts, i.e., its line segments. Its parts exist, so to speak, as potential resultants of division[37]. Divisions can be carried out anywhere along the line. Points on the line exist not actually, but only as potentialities, being in a sense the possible positions where future divisions can be carried out. "The expression divisions can be carried out anywhere along the line" is a fuzzy way of expressing conveying the information that the line is *potentially* dense. And this means not that for any two different points A and B in the line there exists a point C, such that C is after A and before B, but that for any two mental acts of division A and B we can always perform a third C in between the previous two.

The property of sequential or Cauchy completeness as a characteristic of continuity is not something of which Leibniz was fully aware. It appears that whenever he talks about limits he does it for particular cases, where these limits are already known to him[38]. That is to say, he is not fully aware of the property, because if he were, he would have used it as the *a priori* ground for supporting the claim that limits of Cauchy sequences[39] always exist, whether we know them or not. It is correct to say that whenever he refers to continuity as the respecting of limits, he does it by talking about connections between

---

[36] In Chapter III an explanatory model concerning the connection of "spatial betweenness" with its metaphysical correlate via the idea of indirect representation will be presented.

[37] One can easily spot the Aristotelian influence on Leibniz concerning this conception.

[38] When he talks, e.g., of equality as a limit case of inequality, rest as a limit case of motion, etc. (G, III, 52-53.)

[39] For which sequences he would not, of course, have used this terminology.

data and their results, in a way that is very near to what we would call "continuous function talk." But this is not enough, because the existence of continuous connections between data and their results does not imply that limits of Cauchy sequences always exist. What it does imply is that if such limits exist in the set of the data, then they would be mapped onto similar limits in the set of the results.

Sequential or Cauchy completeness is a property of continuity, describable by a universal second-order sentence, according to which every Cauchy sequence is convergent. Such a second-order universal sentence is not to be found anywhere in the Leibnizian writings. If it were, he would have used it; it would have had an extremely rationalistic effect, in the sense that it would tell the empiricist to look for a limit-point wherever appropriate accumulations of relevant points occur. This is not to say that no traces of this property can be found in Leibniz's writings. Let us consider the following passage from a letter of Leibniz to Pierre Varignon written in 1702:

> ... I think I have good reasons for believing that all the different classes of beings whose assemblage forms the universe are, in the ideas of God who knows distinctly their essential gradations, only like so many ordinates of the same curve whose unity does not allow us to place some other ordinates between two of them because that would be a mark of disorder and imperfection. Men are therefore linked with the animals, these with the plants, and the latter directly with the fossils which in their turn are linked with those bodies which the senses and the imagination represent to us as perfectly dead and formless. Now the Law of Continuity demands that when the *essential determinations of one being approximate those of another, as a consequence, all the properties of the former should also gradually approximate those of the latter.* Hence it is necessary that all the orders of natural beings form but a single chain in which different kinds like so many links clasp one another so firmly that it is impossible for the senses and the imagination to fix the exact point where one begins or ends; all the species which border on or dwell, so to speak, in regions of inflection or singularity are bound to be ambiguous and endowed with characters related equally well to neighboring species. Thus, for example, the existence of Zoophytes, or as Buddaeus calls them, Plant-Animals, is nothing freakish, but it is even befitting the order of nature that there should be such. So great is the force of the Principle of Continuity in my philosophy, that I should not be surprised to learn that creatures might be discovered which in respect to several properties, for example, nutrition or reproduction, could pass for either vegetables or animals, and that would upset the commonly accepted rules based on the assumption of a perfect and absolute separation of the different orders of simultaneous creatures that fill the universe. I say I should not be greatly surprised, but I am even persuaded

> that there ought to be such beings which Natural History will some day come to know when it will have studied further that infinity of living beings whose small size hides them from ordinary observation and which are buried in the entrails of the earth and in the abyss of the waters. (W, 186-188; BC, II, 558-559.)

It was necessary to quote this long passage because of its special importance for both, our overall discussion of continuity and our specific discussion of sequential or Cauchy completeness. In light of it, the following comments, concerning, first, the connection between the continuum of real beings and the continuum of beings as they are represented, and, second, the property of sequential or Cauchy completeness, seem appropriate :

(a) Leibniz gives a rather detailed description of what he considers to be the linear continuum of beings as they are represented by us, i.e., as animals, plants, fossils, etc. In this continuum "men are ... linked with the animals, these with the plants, and the latter directly with the fossils, which in their turn are linked with those bodies which the senses and the imagination represent to us as perfectly dead and formless."

(b) As we have said, the continuum of beings as they are represented is in one to one correspondence with the continuum of beings as they really are. Since the only real (i.e., metaphysical) beings are monads, their continuum must be one of the dominant monads of whatever is represented (distinctly, confusedly, or even unconsciously) by us, as an organic whole[40] (an animal, a plant, a man, etc.) "The Law of Continuity demands," says Leibniz, "that when the essential determinations of one being approximate those of another, as a consequence, all the properties of the former should also gradually approximate those of the latter." But that which determines essentially a monad is its representational structure. Beings at the phenomenal level are continuously ordered according to their phenomenal properties (nutrition, reproduction, etc.), which are continuous functions of their "essential determinations", that is, phenomenal properties are continuous functions of characteristics of the representational structures of the monads constitutive of phenomenal beings. Such characteristics are, for example, the relative confusedness and distinctness of the monadic representations, both as wholes and as manifolds of representational parts. Their continuous gradation is responsible for the continuous gradation of the corresponding monads.

(c) The explanation for such a one to one correspondence between the continuum of beings as they really are and as represented is that both are

---

[40] As we have said, for the case of what appears to us perfectly dead and formless, any monad represented as belonging to it must be placed at the very end of the continuum.

different, but isomorphic realizations of one and the same continuum of representables preexisting in the mind of God.

(d) The passage that was quoted above contains very interesting information about Leibniz's conception of what it is to be continuous. We will focus our attention on two points which seem to be relevant to our discussion of the properties of density and sequential or Cauchy completeness. The first concerns Leibniz's description of what he thinks to be the case that obtains near points of "inflection or singularity" in the continuous spectrum of beings[41], as they are represented by us. The second is related to the existential status of the points themselves. Initially Leibniz does not say much about these points. Instead, he talks about regions of "inflection or singularity": "Species which border or dwell" in these regions are "bound to be ambiguous and endowed with characters related equally well to neighboring species." The picture that he most probably had in mind is that of two sequences approaching each other from different directions, so that the difference between their members gets smaller and smaller approaching zero. Then he goes on to talk about the points of "inflection or singularity" themselves, postulating the existence of creatures which correspond exactly to them. He does it for the particular case of creatures which could pass for either vegetables or animals by stating that, he would not be "greatly surprised" if such creatures were discovered and that he is "even persuaded" of their existence. The ground he offers for such a strong statement is the principle of continuity, as the basic architectonic feature of this best of all possible worlds.

It is quite clear that in the passage which was quoted above, density as a characteristic of continuity, is not at the fore. Yet it seems fair to say that Leibniz is not quite aware of it. He seems to think that zoophytes have to exist, because continuity, and more specifically density, dictates that between animals and plants creatures with intermediate properties have been placed. Consider, for example, the following passage from a letter of Leibniz to Louis Bourguet, dated August 5, 1715, where he seems to conjecture that the existence of zoophytes, is dictated by density:

[41] We should note here that Leibniz means to deny that the concept of "species" applies, in any case, to more than a single individual. The following quotation from the *New Essays* is quite revealing:

> In mathematical strictness, the tiniest difference which stops two things from being alike in all respects makes them of 'different species' ... Two physical individuals will never be perfectly of the same species in this manner, because they will never be perfectly alike (NE, 308; G, V, 287-288).

> Of what the perception of plants consists we cannot say; indeed, we do not have any good conception of that of animals. But it is enough to say that the plant has a variety in unity and therefore has a perception; and it is enough that it has a tendency toward new perceptions and therefore appetite, in the general sense that I use these terms. Mr. Swammerdam has supplied observations which show that insects are close to plants with respect to their organs of respiration and that there is a definite order of descent in nature from animals to plants. But perhaps there are other beings between these two. (L, 664: G, III, 581.)

Yet, even if such a creature or creatures did not exist it would be possible for density not to be violated. That is, although Leibniz seems to think otherwise, it is not density which dictates that "perhaps there are other beings between these two", but continuity in the form of sequential or Cauchy completeness.

Moreover in the above quoted passage Leibniz talks about a "definite order of descent in nature from animals to plants" which can be cashed out in terms of a Cauchy sequence since such a sequence can be thought of as a genuine monotone (descending) sequence of animals bounded below by the plants. Such a sequence, if continuity in the form of sequential or Cauchy completeness applies, has to have a limit. Such a limit can only be a zoophyte.

But let us return to our contention that "even if such a creature or creatures did not exist it would be possible for density not to be violated". Let us assume that the chain of beings is isomorphic to the set of real numbers [0,1] -{1/2}, ordered in the usual way. The set [0,1] - {1/2} is the set of real numbers between 0, 1, with 0, 1 included and 1/2 excluded. The real number 1/2 could correspond to a Leibnizian point of "inflection or irregularity" in the chain of beings, with no existing creature to occupy such a point. The set [0,1]-{1/2} is dense, because for any numbers a, b of the set, there is a number c in the set such that $a<c<b$. Rephrasing it for the case of the linear chain of beings, it would mean that although no existing creature or creatures correspond to this point, for any two existing creatures A and B there exists a third creature C, with properties such that C is placed in the chain after A and before B. What is violated in such a case is not the property of density, but that of sequential completeness.

The predominant characteristic of continuity used in the quoted passage from the letter of Leibniz to Varignon is not density, but sequential completeness. Leibniz does not speak of a limit *known* to exist, approached by a sequence or sequences of points in a chain. If he were talking about a known limit, he would not need to use sequential completeness. This sort of completeness is an *existence postulate* of limits of Cauchy sequences and therefore serves as the *a priori* ground for such an existence. Without fully realizing it, Leibniz uses sequential completeness, since he insists that the force of the principle of continuity is so great in his philosophy, that he is "even persuaded" of the existence of the common limit of plants and animals as they approach one another. On the other hand, if he had fully realized the force of the principle of continuity, as involving sequential or Cauchy completeness, he would have insisted explicitly that, at least in the case of linear continua, given *any* two sequences of points *approaching each other* from opposite directions their *common limit exists*. Using the appropriate mathematical jargon, "approaching each other" means that, given the two sequences $a_1, a_2, ..., a_n, ...$, and $b_1, b_2, ..., b_n, ...$, (such that $a_p \leq a_r$ and $b_p \geq b_r$ for every p and r with $p < r$) for every real number $\varepsilon>0$, there exists a natural number m, such that for every natural number k, with $k>m$, we have $|a_k - b_k| < \varepsilon$. It is obvious that both such sequences are Cauchy. Furthermore, we can blend them together in the following sense, $a_1, b_1, a_2, b_2, ..., a_n, b_n, ...$, so that they can be thought of as proper subsequences of one and the same Cauchy sequence. Then the existence of their common limit is translatable into the existence of a unique limit for the sequence that resulted from their blending.

D. THE ROLE OF CONTINUITY IN THE FORMATION OF THE LEIBNIZIAN SYSTEM

The belief that continuity (and the problem of the "labyrinth of the continuum") played a quite important part in the formation of the Leibnizian system is not a novel one. Russell, for example, admits that only reasons of logical priority prevented him from starting his *Critical Exposition of the Philosophy* of *Leibniz* with the question, "How can that which is continuous consist of indivisible elements?"

> We now reach at last the central point of Leibniz's philosophy, the doctrine of extension and continuity. The most distinctive feature of Leibniz's thought is its preoccupation with the "labyrinth of the continuum". To find a thread through this labyrinth was one main purpose of the doctrine of monads-a purpose which, in Leibniz's own opinion, that doctrine completely fulfilled. And the problem of continuity might well be taken, as Mr. Latta takes it (L.21) [42], as the starting point for an exposition of Leibniz: "How can that which is continuous consist of indivisible elements"? To answer this question was, I think, one of the two chief aims of Leibniz's doctrine of substance and of all that is best in his philosophy. That I did not begin with this question, was due to motives of logical priority. (p. 100)

[42] See [92]

With the above question in mind we will explain the role that, we think, continuity of the spatially extended played in the formation of Leibniz's philosophical system. We will consider, as the central point in our discussion, the tension between the idea of this world of ours as consisting in undivided substantial units and the idea of this same world as characterized by infinite divisibility of the spatially extended.

The notion of substance plays a central role in the Leibnizian system. Leibniz, following the traditional route marked out by the Greeks and refined by the Scholastics, was convinced that the ultimate philosophical question was what substance is. His mature answer to this question contains elements which were added, little by little, throughout his entire intellectual life, i.e., it contains elements which were not there when he started contemplating the question. On the other hand, we can trace some elements to the very beginning of his philosophical work.

Such elements played the role of assumptions which were not to be challenged. The first assumption is of a pluralistic, the second of a foundationalist nature. From the very beginning Leibniz was convinced that what is, is not one, but many. On the other hand, such a plurality had to be a plurality of ultimate units, which were to be understood as, among other things, partless, and therefore indivisible (either potentially or actually.) It seems that he was convinced that entities which are partless and therefore indivisible exist and constitute the backbone of what is real. Individual minds were the paradigm of such entities for him[43]. On the other hand his conviction that what is, is not one, but many was grounded on the model of the infinite divisibility of an extended body as indicating the existence of parts. In his *Dissertatio de arte combinatoria* written in 1666 he says:

> 8. Axiom 4. Every body whatsoever has an infinite number of parts; or, as is commonly said, the continuum is infinitely divisible. (L, 74; G, IV, 32.)

We can see Leibniz's philosophical development as the struggle to combine both his pluralist and foundationalist tendencies into one metaphysics, which would accommodate at the same time the idea of the infinite divisibility of what is or appears to be extended. This is not to say that his development was due solely to such a struggle. Logical and metaphysical considerations of a different sort also helped. The fact that we do not care to take sides in the continuing controversy concerning the priority of the logical over the metaphysical foundations of the Leibnizian system is due to two reasons: first, it is beyond the scope of this work and, second, we think the distinction is

---

[43] See G, I, 52-53.

misconceived. Leibniz's mature system was the result of a piecemeal process, involving both logical and metaphysical considerations in its formulation. We see the struggle between the pluralist and foundationalist tendencies as an important part of this process. Leibniz considers his experiential grounds for a pluralist and foundationalist metaphysics to be the undisputed for him fact of the divisibility of the spatially extended and the "clear," but not "distinct" idea we have of an indivisible substance by our internal experience of ourselves as partless and indivisible unities[44]. In other places he ignores the experiential grounds he thinks we have for the idea of the existence of indivisible substances, and makes an attempt to derive such an existence from that of composite ones[45]. He seems to consider the existence of composite substances as something well known and beyond doubt. We think that this shows how certain the divisibility of the extended was for him (which is evidence indicating the existence of composite substances) compared with the internal experience of ourselves (which is evidence indicating the existence of simple-i.e., partless and therefore indivisible-substances.)

The argument seems to run as follows: every being by aggregation (or composite substance) is endowed by some kind of reality. Such reality can only be parasitic upon the reality not of that which is composite, but on the reality of that which is simple. There are beings by aggregation. Concerning

---

[44] Consider, e.g., the following passages:

> I am of opinion that reflection suffices for finding the idea of substance in ourselves, who are substances. (NE, 105; G. V, 96.)

> I believe that we have a clear but not a distinct idea of substance, which comes in my opinion from the fact that we have the internal feeling of it in ourselves, who are substances. (R, 215; G, III, 247.)

> Since I conceive that other beings have also the right to say I, or that it may be said for them, it is by this means that I conceive what is called substance in general. (R, 215; G, VI, 493.)

[45] Consider, e.g., the following passages:

> I believe that where there are only beings by aggregation, there will not even be real beings; for every being by aggregation presupposes beings endowed with a true unity, because it derives its reality only from that of those of which it is composed, so that it will have none at all if each being of which it is composed is again a being by aggregation ... I do not grant that 'there are only aggregates of substances,' and if there are aggregates of substances there must also be genuine substances from which all the aggregates result. (M, 120; G, II, 96.)

> ... what is not truly one being is not truly a being either. (M, 121; G, II, 97.).

the composition of beings by aggregation we can distinguish two possibilities: either such a being is the result of an aggregation of simple substances, or it is the result of an aggregation of other beings by aggregation, which, in their turn, are the results of other beings by aggregation, and so on *ad infinitum*. If the second alternative obtains, the initial being by aggregation, and for that matter, every such being, is devoid of any kind of reality. Simple substances, therefore, exist and are the ultimate constituents of every being by aggregation.

According to this argument, if simple substance did not exist, then, beings by aggregation would not exist either, since their existence would contradict their lack of any sort of reality. For Leibniz, in his mature period, the reality of beings by aggregation was a well-founded phenomenon, based upon the ultimate reality of monads representing one another. From such a point of view, if simple substances did not exist, either there would be no phenomena or no phenomena would be well-founded. But since experience rules out the first alternative, we are left with the second. At this point the question arises : how do we know that phenomena, which can be distinguished from dreams or hallucinations by their coherence, are well-founded? The answer is, "we do not." We need the metaphysician to assure us of their being well-founded.

The Leibnizian argument concerning the derivation of the existence of simple substances from the well-founded phenomenal existence of the compound ones is therefore defective and circular. In order to prove such an existence, we need to presuppose it as the firm basis of well-founded phenomena. Leibniz could have done without such a proof, if he had instead postulated the existence of simple substances. He could have used it, as he often does, as a basic doctrine of his system. We can, therefore, dismiss the argument, without much harm to the system as a whole. We need only accept as an axiom, that composite substances are well-founded phenomena, based upon the reality of partless substantial units representing one another in a pre-established, harmonious way.

As we have seen, the infinite divisibility of the spatially extended was an unquestionable fact for Leibniz, from the very beginning of his philosophical endeavor. The existence of parts in material things was for him strong evidence for the existence of composite substances. Composite substances, on the other hand, had to be based upon the existence of ultimate indivisible substantial units. But how can we combine such an existence with the infinite divisibility of the spatially extended?

Leibniz's early views were rather materialistically oriented. He located in space what he considered, at the time, partless, substantial units. Such partless, substantial units were souls or minds[46]. In his *Paris Notes* we find the following passage, written on February 11, 1676:

> Every mind is of unbounded duration. Every mind also is implanted indissolubly in certain matter. (L, 160; J, 38-40.)

We can also find passages from the period before his journey to Paris, where he describes his views concerning the indivisibility of a mind. He was thinking of it on analogy with a point, which is rather to be expected. After all, since he was committed to the idea of the infinite divisibility of the spatially extended, indivisibility could be thought of by him as a characteristic only of a point; "mind," he says, in an undated letter of his to Duke Johann Friedrich of Braunschweig-Lüneburg, "consists in a point or centre, and is therefore indivisible, incorruptible, immortal ... Mind is a little world, comprised in a point, and consisting of its ideas, as a centre, though indivisible, consists of angles" (R, 123; G, I, 61.)

In an another letter of his to Johann Friedrich, dated May 21, 1671, Leibniz says that:

> ... mind itself consists properly in only a point of space, whereas a body occupies a place... If we give the mind a greater place than a point, it is already a body, and has parts external to each other; it is therefore not intimately present to itself and accordingly cannot reflect on all its parts and actions... But assuming that the mind does consist in a point, it is indivisible and indestructible... I almost think that every body, whether of men or animals, vegetables or minerals, has a kernel of its substance ... (R, 253; G, I, 52-53.)

and he continues:

> If now this kernel of substance, consisting in a physical point (the proximate instrument, and as it were the vehicle, of the soul, which is constituted in a mathematical point), always remains, it matters little whether all gross matter ... is left over. (R, 254; G, I, 54.)

In these passages we find quite interesting traces leading to his middle and later views concerning the distinction between what he called mathematical, metaphysical and physical points[47]. At this early stage of his philosophical development, Leibniz locates souls in spatial points, which he confuses with mathematical points. Such a confusion comes from conflating space as an

---

[46] In a note added by Leibniz in his copy of a letter to Des Bosses, dated April 30, 1709, and probably not included in the original, we find the following frank comment made by Leibniz about that period. "Many years ago, when my philosophy was still too immature, I located souls in points ... " (L, 599; G, II, 372.)

[47] See G, IV, 482-483.

entity in itself with what was taken to be its mathematical representation, i.e., the three dimensional Euclidean continuum. The notion of a metaphysical point, on the other hand, was not present yet in his writings. When he spoke of a physical point, Leibniz was thinking, quite vaguely, about an infinitesimally small, but still extended area, where the kernel of the substance of a body was supposed to be located. What he meant exactly by such a substantial "kernel" is not clear. What is clear, however, is that the idea of a substantial kernel was the precursor of his idea of a corporeal substance existing before the birth or after the death of an organism, so that "all its organs together make only one *physical point* with respect to us" (L, 456; G, IV, 483.)

Let us return to the early view that souls are located in spatial points. Leibniz says in a passage written by him on February 11 1676, and included in his *Paris Notes*:

> ... matter is nothing but a multitude of infinitely small points or of bodies less than any assignable ones, or that there is necessarily an interspersed metaphysical vacuum; this does not conflict with a physical plenum. A metaphysical vacuum is an empty space, however small, yet true and real. A physical plenum is consistent with an assignable metaphysical vacuum. Perhaps it follows from this that matter is divided into perfect points or into all the parts into which it can be divided. (L, 158; J, 30.)

Here Leibniz seems to be saying conflicting things. We will try to reconstruct the picture we think he had in mind at the time. He says that "matter is nothing but a multitude of infinitely small points or of bodies less than any assignable ones, or that there is necessarily an interspersed metaphysical vacuum." He means that matter, in the final analysis, is discrete in such a way that it is constituted of physical points (kernels of substance) of magnitude less than any assignable one. These points are separated by a metaphysical vacuum, which he describes as infinitesimally small empty space. In other words, the magnitude both of these physical points (in which unextended substantial units are located) and of the interspersed metaphysical vacuum is so small that it can be thought of as negligible, compared with the magnitude of the usual physical bodies.

If this is correct, we can distinguish three kinds of spatial magnitude, as related to the picture we are describing. First, souls or minds hold spatial positions, which can be thought of on the analogy of a mathematical point. That is to say, they exist in space occupying a position of magnitude equal to zero. Second, souls or minds are located in physical points which are kernels of substance. In order to describe a kernel of substance Leibniz uses the metaphor "the proximate instrument" and "vehicle of the soul." What is the real (i.e., metaphysical) composition of such a kernel remains obscure. The only thing that can be said about it is that it is dominated by the soul. In a sense, the idea of a kernel of substance is the precursor of the later idea of a whole organism in a state of stupor (equated by Leibniz with the state of the organism before its birth or after its death.) An organism, in such a state, would occupy a physical point and, though diminished, would retain all its essential organs. Physical points, and the empty space that separates them, are extended. They are, that is, of magnitude greater than zero, but so small that it could be considered as negligible. Negligible with respect to what, one wonders. Presumably, Leibniz means with respect to the magnitude of familiar extended wholes, such as rocks, chairs, tables, animals, plants, etc. The magnitude of such composite entities would be the third kind of spatial magnitude in the early Leibnizian metaphysical schema.

We can make more sense of this picture if we consider the magnitude of a physical point or of the interspersed metaphysical vacuum as being of a different order of magnitude from that of the usual physical objects. Using modern mathematical jargon and ideas from the Non-standard Analysis introduced by A. Robinson (see[127]) we could say that for the two orders there is no Archimedean bridge connecting them. A physical point is infinitesimally smaller than any usual physical object. To be of infinitesirnal magnitude would mean, in a sense, to be of a magnitude incomparably smaller than the magnitude of any extended physical object. In other words, if a is the magnitude of a physical point and b is the magnitude of a physical object (a chair, for instance), then a is not equal to zero, but there is no natural number n such that $n \cdot a > b$. In such a case, any extended physical object would consist of infinitely many physical points, such that it is infinitely divisible into extended parts of infinitesimal magnitude. In that sense, a physical plenum would be consistent with the existence of a metaphysical vacuum. The spatial world would be a physical plenum, because vacua of non-infinitesimal magnitude would not exist; the world would consist of extended physical objects which are not infinitesimally small. A metaphysical vacuum would be any empty space of infinitesimal magnitude separating physical points, which would also be of infinitesimal magnitude.

It is perhaps unwarranted to argue that Leibniz had precisely this schema clearly in mind. Yet, such a schema represents the only way we have of making sense of the view Leibniz expresses in the above quoted passages. This is not to say that it is free of problems. We have to face the Leibnizian contention that "perhaps it follows from this that matter is divided into perfect points or into all the parts into which it can be divided." It is not clear how we can think of matter as divided into perfect points. Physical points do not qualify as perfect points, because they are extended. Does Leibniz mean that even physical points are infinitely divided into perfect points? And if so, what is to be a perfect material point? It must be a soul or a mind with materiality as one of its properties. But such an answer, although rather correct, is not a complete one.

We can now locate some of the problems that arise from such an approach. A series of bothersome and disturbing questions are related to it. First, how does extension arise out of unextended units? Second, how can that which is infinitely divisible consist of indivisible elements? How can the continuous be analyzed into a discrete multitude of partless units[48]?

All of these questions were interesting to Leibniz. At the ideal level, we know that given a line segment, we do not change its magnitude by adding a point to it. In his mature period, Leibniz holds that even an infinitude of discrete, unextended points collected together would not suffice to make an extension. The following passage from a letter of Leibniz to Des Bosses, dated April 30, 1709, is quite revealing:

> A point is not a definite part of matter, nor would an infinite number of points gathered into one make an extension. (L, 597; G, II, 370.)

How, then, can that which is spatially extended ultimately consist of unextended and partless units? *If to be in space is to be extended and to be extended is to be infinitely divisible (and therefore to have parts) how can partless substantial units be thought of as being in space?* The metaphysical schema we find in *Paris Notes*, seen in light of the above questions, was not satisfactory even to Leibniz himself. In the same *Notes* we find passages in which he betrays his lack of confidence in the solutions he had proposed. It seems that he considered them as only tentative, first approximations to the unraveling of the labyrinth of the composition of the continuum:

> The whole labyrinth about the composition of the continuum must be unraveled as rigorously as possible ... We must see whether it can be demonstrated that there is something infinitely small yet not indivisible; from the existence of such a being there follow wonderful things about the infinite; namely if we assume creatures of another, infinitely small world, we will be infinite in comparison with them. (L, 159; J, 34-36.)

To unravel the labyrinth of the continuum was a matter of the first priority for Leibniz. He thought he had a solution, namely, to distinguish the infinitely

---

48 Another important question is, how can matter or that which is merely passive be so reduced to that which is real and active, and what is it that we say when we insist that a soul has materiality as one of its properties?

small-the existence of which had not been demonstrated, and was an open problem for him-from the usual extended objects, such as chairs, tables, rocks, etc. Such typical extended objects were supposed to be infinitely larger than these infinitesimally small, but still extended somethings. Leibniz was not sure about the existence of these infinitesimally small beings, which were to occupy infinitesimally small spatially extended positions. Later he dropped the claim that they exist, insisting that infinitesimals as well as actual infinites, as referring to united wholes in nature, were useful fictions of the mind, and that their usefulness was restricted to the domain of the ideal, where the science of mathematics predominantly belongs.

Concerning infitesimals he insisted that it is not only the case that "we cannot penetrate to them" but, moreover, they are not even possible in nature because if they were possible they would exist. The following passage from a letter to Johann Bernoulli dated November 18, 1698, is quite illuminating:

> As concerns infinitesinal terms, it seams to me not only that we cannot penetrate to them but that there are none in nature, that is they are not possible. Otherwise, as I have already said, I admit that if I could concede their possibility, I should concede their being. (L, 511; GM, III, 551).

Leibniz was already convinced, at the time, that infinitesimals as well as infinities (that is the infinitely small and the infinitely big) were not of this world. They were imaginary. They were, useful fictions of the mind. Let us have a look at the following passage, written in 1698, in which Leibniz suggests that the infinites and infinitesimals of his calculus were:

> ... imaginary, yet fitted for determining reals, as are imaginary roots. They have their place in the ideal reasons by which things are ruled as by laws, even though they have no existence in the parts of matter.

As for the "real infinite," he continues, it "is perhaps the absolute itself, which is not made up of parts, but which includes beings having parts, in an eminent way and in proportion to the degree of their perfection" (L, 511; GM, III, 499-500.) This is not to say that the notion of "actual infinity", as referring, for example, to a multitude of monads, was altogether refuted by Leibniz. What he wanted to deny was the existence of an infinite number referring to a composite whole, which would be considered as a *true* substance by itself, as opposed to an infinite aggregate of parts. Consider, e.g., the following passages, the first from an undated letter to Simon Foucher and the second from the *New Essays on Human Understanding*:

> I am so much in favor of the actual infinite that, instead of admitting that nature abhors it, as is commonly said, I hold that nature affects it everywhere, in order the better to mark the perfections of its author. So I believe that there

is no part of matter which is not, I do not say divisible, but actually divided; and consequently the least particle must be regarded as a world full of an infinity of different creatures. (R, 109; G, I, 416.)

It is perfectly correct to say that there is an infinity of things, i.e., that there are always more of them than one can specify. But it is easy to demonstrate that there is no infinite number, nor any infinite line or other infinite quantity, if these are taken to be genuine wholes. (NE, 157; G, V, 144.)

In the first of the two quoted passages Leibniz talks about the actual infinity of an aggregate as a consequence of the fact that matter is not simply infinitely divisible but actually infinitely divided. In the second he repeats that there is in actuality an infinity of things but he is quick to add that such infinities do not refer to or, even better, are not "genuine wholes".

Genuine wholes do not exist in nature, where the term "genuine" has to be understood, in Leibniz's terminology, as signifying real substances, the paradigm for a real substance in nature being only that of a substantial partless unit. Leibniz came to realize that such a distinction between infinitesimally small beings and more familiar extended beings such as chairs, tables, etc. could not alone allow him to unravel the labyrinth of the continuum. Whether there are infinitesimally small beings or not, the fact remains that they must also be infinitely divisible, since they are extended, and therefore have parts *ad infinitum*. What were supposed to be the ultimate constituents of these infinitesimally small beings?

Substantial, partless units of course! But the puzzle remains, as long as substantial partless units are assumed to occupy unextended spatial positions. The questions supposed to be answered were, first, how the continuous consists of discrete partless units, and, second, how the extended-whether infinitesimally small or not-can be composed of unextended entities. Leibniz found his way out of the puzzle later. As Russell puts it:

It must have been soon apparent to Leibniz that this doctrine did not solve the difficulties of the point and the instant, or afford a consistent theory of substance. And so we find, in his early published accounts of the doctrine of monads, a third kind of point added to the above two [i.e., to the mathematical and the physical point], namely the metaphysical point, while the mathematical point is no longer that in which the soul consists, but only its point of view. ([129], pp. 123-124.)

Leibniz's solution to the puzzle was to free substantial units (simple substances, or monads) from their spatially extended prison cells. He stopped locating them in spatial points. This new metaphysical position was the right one to adopt. It provided him both with a solution to the problem of the composition of the continuum and with a consistent theory of substance.

Extended objects can be divided *ad infinitum*, but this infinite divisibility takes place only at the phenomenal level. The metaphysical level forms the actual basis for these phenomenal divisions, because it consists of already separated partless substantial units representing one another as if they were *continuously* positioned in space. They are "positioned in space" not as extended wholes, but as the ultimate constituents of these wholes. According to this new approach, it is meaningless to ask which are the ultimate indivisible units, as long as the question is asked at the level of the phenomena. The question must be asked at the metaphysical level, and there the answer is clear and obvious. The ultimate indivisible units are the simple substances, which are metaphysical atoms of some sort, or non-spatial, non-temporal points representing one another as if they were in space and time. Let us see how Leibniz describes the move of freeing himself "from bondage to Aristotle" and of freeing simple substances from their spatially unextended confinements in a passage from his essay "A New System of the Nature and the Communication of Substances as well as the Union between the Soul and the Body" which appeared in the well-established Paris journal *Journal des savans* (June 27, 1695):

> At first, after freeing myself from bondage to Aristotle I accepted the void and the atoms, for it is these that best satisfy the imagination. But in turning back to them after much thought, I perceived that it is impossible to find *the principles of a true unity* in matter alone or in what is merely passive, since everything in it is but a collection or aggregation of parts to infinity. Now a multitude can derive its reality only from *the true unities*, which have some other origin and are entirely different from points, for it is certain that the continuum cannot be compounded of points. To find these *real unities*, therefore, I was forced to have recourse to a formal atom, since a material being cannot be at the same time material and perfectly indivisible, or endowed with true unity (L, 454; G, IV, 478[49].)

[49] This passage, as it has been already said, is from the original essay published in the *Journal* des *savants*, 27 June 1695, under the title "Système nouveau de la nature et de la communication des substance aussi bien que de l'union qu'il y a entre l' âme et les corps". In Leibniz's final version the last two sentences were altered as follows: "Now a multitude can derive its reality only from true unities which have some other origin and are entirely different from mathematical points, these being merely the extremities of what is extended, and modifications of which it is certain the continuum cannot be composed. To find the real unities, therefore, I was forced to have recourse to a *real and animated point*, so to speak, or an atom of substance which must include a certain active form to make a complete being."

At this stage of his development, Leibniz distinguishes three kinds of points, metaphysical, mathematical, and physical. In a passage from the same essay he says:

> 11. Furthermore by means of the soul or form there is a true unity corresponding to what is called "I" in us. Such a unity could not occur in artificial machines or in a simple mass of matter, however organized it may be. For such a mass can be compared only to an army or a herd, or to a pond full of fish, or a watch made of springs and wheels. If there were no true substantial unities, however, there would be nothing substantial or real in the collection. It was this that forced Cordemoi to abandon Descartes and to support the Democritean theory of atoms in order to find a true unity in them. But *material atoms* are contrary to reason, besides being still further composed of parts ... It is only *atoms* of *substance*, that is to say, real unities that are absolutely destitute of parts, which are the sources of action and the absolute first principles out of which things are compounded, and, as it were, the ultimate elements in the analysis of substance. One could call them *metaphysical points*. They have something vital, and a kind of *perception* and mathematical points are the *points of view* from which they express the universe. But when a corporeal substance is contracted, all its organs together make one *physical point* with respect to us. Physical points are thus indivisible in appearance only, while mathematical points are exact but nothing but modalities. It is only *metaphysical* points, or points of substance, constituted by forms or souls, which are exact and real, and without them there would be nothing real, since there could be no multitude without true unities. (L, 456-457; G, IV, 482-483.)

Metaphysical points are the simple substances (monads.) They are the source of every action and the ultimate constituents of everything compounded. They are characterized by their particular representational structure and they are endowed with something vital, i.e., the non-voluntary ability to move continuously from representational state to representational state (appetition), each of which is a momentary static perspective view of the universe. No transient interaction between simple substances can take place and so they are completely closed and self-sufficient. They are living mirrors of the whole universe, which itself – metaphysically speaking – is only a manifold of such living mirrors. Each such mirror represents the whole universe perspectively, i.e., from a particular point of view. Leibniz calls these particular points of view mathematical points, still confusing a mathematical with a spatial point (if such a thing has any meaning at all), because of structural similarities between them. According to Russell, at this stage of Leibniz's development:

> ... space and the mathematical point retained more reality than was to be wished, and accordingly both the expression "metaphysical points," and the assertion that mathematical points are the points of view of substances disappear after 1695. After this time, he still speaks of points of view, and always explains them on the analogy of spatial points from which the world is, as it were, seen in perspective (G, II, 438; III, 357). But he insists that this is *only* an analogy, without, however, telling us to what it is analogous. ([128], p. 124.)

Finally, physical points are only points in appearance. They do appear as points to us because we cannot distinguish finer details in them and see them as partless. That is, we comprehend them on the analogy of a point, which is an abstract entity. Physical points are extended, but are of very small size. Yet such a size should not be thought of as of a different order (i.e., of infinitesimal order) compared with the size of familiar extended objects, such as tables, chairs, etc. For Leibniz in this period, there are no beings of infinitesimal magnitude. Physical points appear to us as such by our *minimum perceptibile*. It is, perhaps, the case that other subtler animals with a finer such minimum could distiguish details inside a physical point (as we cannot.) This is not to say that we do not represent such details. We do represent them, but confusedly, so that we have only the distinct representation of the whole as a partless point. Optical instruments, such as a microscope, could enhance our ability to distinguish some finer details in it and so *distinctly represent* it as an extended whole. Whatever the case, it remains that a physical point is something which belongs to the realm of the phenomena, it is characterized by extendedness and is therefore infinitely divisible. "Physical points are thus indivisible in appearrence only" says Leibniz.

Leibniz's solution to the problem of reconciling infinite divisibility of the spatially extended with the ultimate reality of indivisible substantial units was, as we have already said, to free these units from their spatially unextended prison cells. This move resulted in a radical transformation of the Leibnizian system. Its most notable result, among other things, was a solution to the problem of the labyrinth of the continuum, which we will examine in detail in the next chapter. We can distinguish also some other profound consequences of this move, as related to Leibniz's metaphysics of space and of time.

In his early writings, Leibniz treats space as something real. This view is clearly visible in a letter (dated April 20/30, 1669), to Jacob Thomasius, one of the most influential of Leibniz's teachers at Leipzig:

> For the rest, I have proved that figure is a substance, or rather that space is a substance and figure something *substantive*, because all science deals with substance, and it cannot be denied that geometry is a science ... Geometry is

> thus a true science, and, Aristotle not to the contrary, its subject, which is space, is a substance. (L, 98; G, I, 21; IV, 168-169.)

Space was real in the sense that it was an absolute container, existing independently of its contents, which could only be real if they were contained in it. Spatial relations in that sense were parasitic upon the pre-existence, first, of space and, second, of the objects in space. Space itself would play the role of the relational bridge and the objects in space the role of the appropriate spatial relata. By adopting this position, Leibniz was once again hopelessly entangled in an uneven fight with the problem of the composition of the continuum. How could indivisible substantial units be in space, if to be in space is to be extended and if to be extended implies to be divisible?

It seems that taking indivisible substantial units out of space and making them the only true reality was for Leibniz the only possible viable alternative. At least he would not have to struggle any longer with the question, what it means to be positioned in space, without being extended. But this move had much more important metaphysical repercussions. Space could no longer be considered real. It seems that Leibniz adopted the following line of thought: if there are no real things to be contained in space (and so situated), there must be no real container either. The statement "there are no real things to be contained" should not be taken to imply "there are no real things." The new Leibnizian doctrine was that *there are real things but they are not contained in space*. Space became an ideal entity prior, but also in a sense conforming, to the *phenomenal* relations, which describe the appearance of a simultaneous arrangement of the partless substantial units, as based upon their harmonious mutual representation. The initial Leibnizian thesis of space as an *absolute container* was replaced by a *relational theory* of *space*. Space could now be thought of as an abstract entity, representing, at the level of the ideal, the manifold of simultaneous phenomenal relations, governing the phenomenal mutual arrangement of monads.

But why did spatial relations have to be phenomenal? It seems that Leibniz could not accommodate spatial relations as real because real space would then enter his system through the back door, in the form of a composite real entity, namely, as the manifold of simultaneous real spatial relations of all existing monads. This would lead to a situation structurally similar to the one he was faced with when he assumed that space was a real, absolute container. He would again be faced with the question of how indivisible substantial units could "be" in space, if to "be" in space is to be extended and therefore divisible. Spatial relations had, therefore, to be phenomenal. But if spatial relations cannot be real, and therefore cannot exist over and above their relata (as do real bridges connecting otherwise disconnected areas), their phenomenality should be cashed out in terms of the metaphor of containment. Spatial relations, in other words, had to be analyzed and reduced into predicates contained in the subjects so related. Such a view was in complete agreement with the more general doctrine Leibniz was committed to, namely, the thesis that relations at the phenomenal level were reducible to relations at the real level (i.e., relations among monads), which in their turn were reducible to predicates contained respectively in the related subjects.

One might wonder whether it was the move of taking indivisible substantial units out of space that forced Leibniz to adopt the more general doctrine of the non-reality of relations, and consequently to accept the thesis that they are parasitic upon the containment of relevant predicates in the corresponding related subjects, or whether the inference went the other way, that is, from first accepting the more general thesis, and then adopting its implication, taking the basic units out of space. The answer is difficult. The thesis of the containment of the predicates in the subject was in the air well before Leibniz. In a sense, its origin was Aristotelian. Our view is that quite, possibly, Leibniz was aware of and sympathetic to it, because such a metaphysical doctrine had its logical counterpart, towards which he was already inclined[50]. Whether he was also already inclined toward an analysis of relations involving only predicates as their ultimate constituents is not so clear. In any case, it remains that Leibniz, by taking indivisible substantial units out of space, came remarkably near to the idea of the non-reality of relations in general, toward which he was perhaps already well-disposed.

The question with which the discussion began was that of how that which is continuous can consist of indivisible elements. We also asked how the spatially extended (which, according to Leibniz, is continuous) can be composed of unextended ultimate substantial units. Both questions were very important to Leibniz, given that he wanted to keep both the infinite divisibility of the spatially extended and the idea of reality as ultimately consisting of partless and indivisible substantial units. His initial metaphysical schema had as its basic ingredient the doctrine that indivisible substantial units occupy (or are positioned on) spatial points. But he came to realize later that spatial points cannot make up extension. To exist in space is to be already extended, and to be extended is to be infinitely divisible. Additionally, faced with the puzzle of the composition of the continuum, he was finally convinced that, we think, what is continuously extended cannot be composed of indivisible substantial

---

50 In his *Dissertatio de arte combinatoria,* for instance, he talks about *derived* and *first* terms, taking the derived ones to be composed out of first terms by *containing* them. (G, IV, 66.)

units. So in order to solve those puzzles he was forced to adopt the view that indivisible substantial units are not spatially situated. Space could not be real and extension had to be a well-founded phenomenon. In the next chapter we will examine the basic ingredients of his solution, as related to the metaphysical foundations of space, time, and the spatio-temporally extended.

CHAPTER III

SPACE, TIME, THE SPATIO-TEMPORAL, AND MONADIC REALITY

A. SPACE, THE SPATIALLY EXTENDED, AND ITS METAPHYSICAL CORRELATE

It is accepted by quite a few Leibniz scholars[1] that one can distinguish three levels in Leibniz's metaphysics. They are the levels of the real, the phenomenal, and of the ideal. These three levels and their interplay constitute the backbone of the system. The tantalizing problem of the composition of the continuum contributed to the formation of such a tripartite system and gave it its special flavor. We will examine the system as it refers to space, time, the spatio-temporally extended, and its metaphysical counterpart. We will also spell out what we consider to be the full Leibnizian answer to the above problem, i.e., what, he thought to be the way out of the labyrinth of the continuum.

We will start our discussion with an examination of what Leibniz thought space to be. As we have seen in the previous chapter, Leibniz in his early writings was committed to the idea of space as a real entity. Space was the absolute container of things contained. He was also committed to the idea of the infinite divisibilty of both the container and the things contained, insofar as these things were extended objects. Infinite divisibility of extended objects, not only as a mere theoretically conceived potentiality, but also as a possibility of actual subdivisions carried out *ad infinitum* was strong evidence, Leibniz thought, for the correctness of the metaphysical doctrine that everything extended is not a real unity, but a plurality of real unities; an *infinite* plurality, in fact, because of the unbounded possibility of carrying out finer and finer subdivisions of that which is extended. But if that which is extended cannot be a unity, what sorts of things are the ultimate substantial indivisible components of the extended compounds and where are they situated in space, if they are? That there must be such indivisible substantial units was a philosophical doctrine of a foundationalist nature, a kind of indispensable prerequisite-and therefore quite dear-to every pluralistic system in the history of philosophy. What was composite had to be composed of what was simple. The non-

[1] See, e.g., [31], [53], [68], [103], [154].

existence of simples would amount to an infinite regress, undesirable to philosophers who deplored the idea of the non-existence of a firm ultimate ground upon which they could base their analysis of substance.

For Leibniz, the internal experience of ourselves as unitary indivisible entities was strong evidence for the existence of indivisible substantial units. Souls or minds were therefore for him, from the very beginning of his philosophical endeavor, the paradigmatic cases of such indivisible substantial units. As partless and therefore indivisible, they could not be extended, but they yet had to be positioned in space. It seems that Leibniz at this time believed that everything in this world deserving of the characterization "real" had to be positioned in space. But because whatever was positioned in space as an extended object had to be infinitely divisible, indivisible substantial units could not be so positioned. What was the alternative? Mathematics[2] was already developed enough to make room for the notion of an ideal entity with the property of non-extension as its basic characteristic. Such a notion was that of a *mathematical point*. So Leibniz located his indivisible substantial units in spatial points, which he conceived on the model of mathematical points. Spatial points were conceived on the one hand as real (or as modifications of the real), because they belonged to what was considered as real, namely space, and on the other as completely isomorphic to those ideal entities which were called mathematical points. It seems that such a complete isomorphism was responsible for Leibniz's initial confusion between a spatial and a mathematical point[3].

By adopting such a position, Lebiniz was helplessly entangled in a maze of problems related to the composition of the continuum. It was unclear, e.g., how the continuously extended could be composed of what is unextended, and how space as a real extensive container could be composed of unextended entities, i.e., spatial points conceived as unextended on the analogy of mathematical points.[4] The Leibnizian reaction to these problems was, we think, not to answer, but to avoid them, by changing the metaphysics of his system appropriately. He stopped locating his indivisible substantial units in space, while maintaining the infinite divisibility of the spatially extended. Indivisible substantial units, though no longer spatially situated, were to be the only real inhabitants of this world. On the other hand their shadows, so to speak, were to be the only inhabitants of the world of phenomena.The spatio-temporal universe of the extended objects became a coherent panorama, a *bene*

---

2 Especially Euclidean geometry, which was thought to be the science of space.

3 See G, I, 52; G, I, 54.

4 Since space was conceived on the analogy of a three-dimensional Euclidean continuum.

fundatum phenomenon, and space and time themselves became not real, but ideal entities.

We will start by examining Leibniz's contention that space in itself is an ideal entity. In his *Correspondence* with Clarke and in his "Fifth Paper" we find the following important passage:

> 33. Since space in itself is an ideal thing, like time; space out of the world must needs be imaginary, as the schoolmen themselves have acknowledged. The case is the same with empty space within the world, which I take also to be imaginary, for the reasons before alleged. (A, 64; G, VII, 396.)

In this passage Leibniz insists that "space in itself" is an ideal entity and "space out of the world", as well as "empty space within the world", are imaginary. By "space out of the world" Leibniz means space as an absolute container, the existence of which he does not accept. Similarly, by saying that "empty space within the world" is "imaginary" he denies the existence of vacua in what we perceive as the material universe. Such a universe is an infinite plenum, both in the sense of containing an infinitude of objects, in a way that no vacua exist, and in the sense of being unbounded. It cannot be finite for the following reason[5]: a finite material universe would not be agreeable to the wisdom of God, who decided to create the best of all possible worlds. Finitude would be a limitation and since any limitation is a sign of imperfection, the best of all possible worlds could neither be *really* finite nor be represented as a finite material universe. Space as an ideal entity, in accordance with the infinitude or unboundedness of what appears to be the material universe, had to be infinite or unbounded also.

The question now arises of determining what the ideality of space meant to Leibniz. At least one thing appears certain, namely, that the ideality of space meant that it is a concept which does not have a real entity as a referent. On the other hand, it is a concept which conforms to the sensible phenomena in a way similar to that in which mathematical concepts like number, extension, figure, magnitude, etc. conform to the same sensible phenomena. Such concepts constitute the backbone of our theoretical framework, which in its turn serves as the tool for both recognizing and explaining what we take the general features of the world of phenomena to be.

---

[5] According to Leibniz (*Leibniz-Clarke Correspondence*, Leibniz's "Fifth Paper"):

> Absolutely speaking, it appears that God can make the material universe finite in extension, but the contrary appears more agreeable to his wisdom. (A, 64; G, VII, 396.)

According to Leibniz, extension and, therefore, space are abstractions from that which is perceived as extended and "the extended is a continuum whose parts are coexistent". The following passage from the *New Essays* is quite illuminating:

> Extension is an abstraction from the extended, and the extended is a continuum whose parts are coexistent, i.e., exist at the same time. (NE, 149; G, V, 136.)

In the following two passages (the first from a letter of Leibniz to De Volder, dated June 30, 1704 and the second from a letter to Des Bosses, dated January 13, 1716,) Leibniz argues, in analogy with numbers, that extension would not "remain if things were removed".

> And, as in counting, the number is not a substance without the things counted, so neither is a mathematical body or extension without active or passive entities or motion. (L, 536; G, II, 268.)

> That extension would remain if monads were removed I hold to be no more true than that number would remain if things were removed. (R, 240; G, II, 510.)

To apply this analogy, as a particular number is an abstraction from the things numbered, so a particular extension is an abstraction from the things so extended. Moving a bit further up the ladder of abstraction, space is a concept which incorporates all the concepts of particular extensions, without itself being a particular extension, in the same way that the concept of the totality of numbers incorporates all the concepts of particular numbers, without itself being a particular number. Number, extension, space, etc., are abstractions not in the sense that they are epistemologically posterior to, and derivable from, our experiential raw material, but in the sense that they are abstract ideal entities which constitute in an *a priori* manner the theoretical backbone of our scientific image of the world.

The interesting thing is that although such concepts are not representations of particular existents, they conform to the sensible phenomena. Such conformity is due to a faithful correspondence pre-established by God between not only what is real and what is phenomenal, but also between all three levels of the Leibnizian metaphysical system. According to Leibniz, the understanding, although aided by the senses, has by itself the capacity to recognize the general features of the world in terms of innate mathematical and metaphysical concepts. This is not to say that the mind is aware of all the innate ideas or the eternal truths it contains. Their recognition and appreciation involves procedures which can be wrongly thought of as bringing *to us* knowledge not already *in us*. The truth, says Leibniz, who here follows Plato in some sense, is that these procedures can be described as revealing concepts or truths about them which were already in us in an unconscious manner. In a passage in the *New Essays* Leibniz categorically asserts that "truths about numbers are in us."

> ... I quite agree that we learn innate ideas and innate truths, whether by paying heed to their source or by verifying them through experience... I cannot accept the proposition that *whatever is learned is not innate*. The truths about numbers are in us; but still we learn them whether by drawing them from their source, in which case one learns them through demonstrative reason (which shows that they are innate), or by testing them with examples, as common arithmeticians do. (NE, 85, G, V, 70-71.)

If we were to extend this discussion so as to include the concept of space and truths about it in what we consider as innate, we could say that ideality of space had for Leibniz two main components: first, space was considered as a concept which in no way corresponded to a real entity; second, it was thought to be an innate idea, i.e., an idea which we are mistaken in thinking is formed *a posteriori*, as an abstraction from what we experience as extended. Space is an ideal entity of a special sort for Leibniz. As the subject of the mathematical discipline of Euclidean geometry, space can be considered as belonging to a special class of concepts, namely the mathematical ones. Euclidean geometry is not an empirical science in the sense that once we have established its axioms we can start unraveling truths about its subject matter, using only ideal mathematical procedures, i.e., ones based upon deductive reasoning. An obvious objection could be raised about the status of the initial axioms, for they seem to have an empirical grounding. Leibniz might answer this objection with an analogy ready to hand. He could admit that the axioms have an empirical grounding, but deny that this makes Euclidean geometry an empirical science; for the empirical grounding is due to the fact that mathematical concepts conform to what is empirically ascertainable, because the "nature of things and the nature of the mind work together[6]." After all, he might say, even the concept of number has an empirical grounding. The initial axioms of arithmetic seem to have such a grounding too. But this does not make arithmetic an empirical science. That Leibniz was considering arithmetic and geometry on a par as sciences whose propositions were about innate ideas is beyond doubt. The following passage from the *New Essays* makes this clear:

---

[6] NE, 84; G, V, 70.

PHIL. What do you say, sir, to this challenge which a friend of mine has offered? If anyone can find a proposition whose ideas are innate, let him name it to me (he says); he could not please me more.
THEO. I would name to him the propositions of Arithmetic and Geometry, which are all of that nature; and among necessary truths no other kind is to be found.
PHIL. Many people would find that strange. Can we really say that the deepest and most difficult sciences are innate?
THEO. The actual knowledge of them is not innate. What is innate is what might be called the potential knowledge of them. (NE, 86; G, V, 71-72.)

In this passage, Leibniz makes clear that two are the main characteristics of Arithmetic and geometry. First, he stresses the point that the propositions of them are the only necessary truths to be found, being propositions "whose ideas are innate". Second, he is quick to add that only the potential knowledge of these propositions is innate, while their actual knowledge, presumably being dependent upon the circumstances of the particular procedures of actualization, exists in us as a mere potentiality.

Leibniz defines space (in juxtaposition to time) in numerous places in his writings. In the following short passage from a letter of his to Des Bosses, dated June 16, 1712, he states explicitly that the order of coexistence of the phenomena is what constitutes the backbone of space and, respectively, the order of succession of the phenomena is what constitutes the backbone of time. More accurately, space and time are, according to him, these respective orders. He says:

... space is the order of coexisting phenomena, as time is the order of successive phenomena. (L, 604; G, II, 450.)

In a later occasion (*The Leibniz-Clarke Correspondence*, Leibniz's "Fifth Paper") Leibniz insist that "...space is nothing else but an order of the existence of things observed as existing together." (A,63; G,VII, 395). In a letter to Nicolas Remond dated March 14, 1714, he insists that "the source of our difficulties with the composition of the continuum" is connected with, or stems from, the fact that we conceive matter and space as substances. He continues defining once again space and time as the corresponding orders of "coexistence" and "of existence which is not simultaneous":

The source of our difficulties with the composition of the continuum comes from the fact that we think of matter and space as substances, whereas in themselves material things are merely well-regulated phenomena, and *space is exactly the same as the order of coexistence, as time is the order of existence which is not simultaneous*. (L. 656; G, III, 612.)

So space, defined as "the order of coexistence," is an ideal entity, an abstraction from the coexisting phenomena themselves. It belongs to a set of entities which are characterized by uniformity and contain no variety. "Things which are characterized by uniformity and contain no variety[7]," Leibniz says, "are never anything but abstractions, like time, space, and the other entities of pure mathematics" (NE, 110; G, V, 100.) Once again, we should understand that the term "abstraction," as used by Leibniz here, does not signify some kind of extrapolation from what is empirically gathered. Leibniz does not mean that we come to have the concept of space, of time, and of other entities of pure mathematics as generalizations out of our raw material of experience. We always have these ideas in our minds, but we are not aware of them. Experience plays the role of the Socratic midwife by helping to bring about at the level of consciousness what was always hidden in the mind. According to Leibniz, this is the reason we can contemplate and find truths about these concepts without having to appeal immediately to what we would call the world of well-founded phenomena. Such truths already exist in the mind, but until we bring them to consciousness through certain mental procedures, we are simply not aware of them[8]. In other words, before we are actual knowers of eternal truths, we are potential knowers of them, in the sense that we carry them in us in an unconscious manner. So abstract concepts such as space, time, number, etc. are already in the mind, and because the "nature of things and the nature of the mind work together" (NE, 84; G, V, 70), when they are brought to consciousness as inductive generalizations from experience, they play the pivotal role in the recognition of the general features of the world of phenomena.

Leibniz, in a passage taken from his "Note on Foucher's Objection" which appeared in the April 1696 edition of the *Journal des savans,* and was a reply to Simon Foucher's letter (which also appeared in the same journal in its September 1695 edition and contained a number of objections to Leibniz's *New System of the Nature and of the Union of the Soul and Body*), maintains that extension and space are relations or orders of existence with no firm bases of composition:

[7] Leibniz means by "uniformity" and "no variety" the basic characteristics of abstract concepts which do not involve specifics.

[8] It is correct to say that there are places where Leibniz does not mention the innateness of those concepts. He appears to be saying that we come to form them by abstracting from the world of experience. See, e.g., A, 69; G, VII, 400. But this is misleading. What he describes there is an experiential procedure for coming to know what is already in us.

> Extension or space, and the surfaces, lines, and points that can be conceived in it, are nothing but relations of order or orders of coexistence, both with regard to that which actually exists and as regards the possible thing that might be put in place of that which exists. Thus they have no ultimate component elements ... (AG, 146, La, 329; G, IV, 491.)

Space, in other words, is an ideal entity which we come to think of as enfolding what appears to us as extended. Spatial relations, in the abstract sense, are extrinsic denominations, quite independent of their possible phenomenal relata. Space can be thought of not as the mere aggregate of all ideal spatial relations-i.e., spatial relations as they are conceived in the abstract framework of Euclidean geometry-but as the ideal entity prior to them, which provide the framework for their mathematical study. Space, so viewed, is a whole prior to its parts and prior to its divisions. Parts of space are the resultants of actualized potential mental acts of division. Such acts, once carried out, result in a mental entity with all the characteristics of space, modified only to the extent that, so to speak, planes, curved surfaces, straight lines, curves, and points are the resultant scars of the acts performed inside the previously immaculate and uniform universe of Euclidean geometry. Such scars are either zero or one or two-dimensional entities. They do not qualify as parts of space. Parts of space are the three-dimensional resultants of mental acts of division, with curved surfaces or planes as their boundaries. Straight lines or curves are the boundaries of the common intersections of curved surfaces or planes, and points the boundaries or the common intersections of straight lines or curves. The scars left over by a mental act of division of space are not parts of space, but extremities; and as referring to the world of phenomena mere modalities. Such *scars* can be left over by an act of division *anywhere* in the body[9] of space. Yet space as an ideal entity has no such scars. Parts of space as ideal entities have a potential status which can be transformed into an actual one by a mental act of division[10].

Leibniz considers points as extremities or modalities. The following passage, from a letter of Leibniz to Des Bosses dated February 8, 1708, is quite illuminating:

> Position is, without doubt, nothing but a mode of a thing, like priority or posteriority. A mathematical point itself is nothing but a mode, namely an extremity. And thus when two bodies are conceived as touching, so that two mathematical points are joined, they do not make a new position or whole, which would be greater than either part, since the conjunction of two

[9] Assuming that we could be allowed catachristically to use such an expression.

[10] The Aristotelian influence on Leibniz is quite apparent here.

> extremities is not greater than one extremity, any more than two perfect darknesses are darker than one. (R, 250-251; G, II, 347-348.)

We can find nowhere in Leibniz's writings a similar commitment with respect to one or two-dimensional ideal entities. That is not to say we find a different commitment either. This is a definite defect of his system. It is a defect, however, which can be understood if one is aware that points were his paradigm of unextendedness. Most of the time Leibniz was struggling with the problem of the composition of the extended out of unextended points. So he never became fully conscious of the fact that the following three problems were basically identical:

(a) How can we build up the linear continuum out of points (i.e., out of zero-dimensional entities)

(b) How can we build up the two-dimensional continuum out of one-dimensional continua?

(c) How can we build up the three-dimensional continuum out of two-dimensional continua?

All three problems are special cases of the more general problem of:

(d) How can we build up an nth-dimensional continuum out of n-1th - dimensional continua, where n is a natural number greater than or equal to one?

Had Leibniz been conscious of the problem in its generality, he would have considered, in the case of space, all the two-dimensional entities as extremities. In the framework of a three-dimensional ideal entity, two-dimensional ideal entities are considered to be unextended, simply because extension in this framework amounts to being three-dimensional. To put it metaphysically, if Leibniz had assumed that his indivisible substantial units were to be conceived on the model of indivisible two-dimensional objects, he would be faced with exactly the same problem of the composition of the continuum; he would be trying to build up a three-dimensional composite continuous substance of two-dimensional indivisible substantial units. It is true, of course, that if he could solve the problem at the level of the linear continuum as composed of points, he could also solve the problem all the way up the ladder of the nth-dimensional continua, where n is a natural number greater than or equal to one.

Leibniz holds that space is infinitely divisible in a very Aristotelian sense. This divisibility is cashed out in terms of the Aristotelian notion of potentiality. Space is an ideal entity having as its basic characteristics a three-dimensional uniformity paired with continuity in the sense of uninterruptedness. Uninterruptedness is a basic characteristic of space. As an ideal entity space does not consist of pieces or bits of extension, i.e., of finite or even infinite three-dimensional regions separated from one another by two-dimensional boundaries. Space is prior to such pieces or bits of extension, in the sense that these are only potentially there. It can be mentally divided anywhere throughout what we consider as its body, but such a mental act is the equivalent of making actual a partition of space, which, prior to the act, had the status of a mere potentiality.

Continuity of space, as an ideal entity, meant the following two things to Leibniz: (a) uninterruptedness and (b) potential divisibility of the whole, anywhere throughout what we conceive as its body. The first characteristic was the one Leibniz appreciated most, thinking it the basic characteristic of all genuine continua[11]. The second characteristic could be thought of as implying at least density in the following sense: between any two actualized points on an actualized straight line a third point can be brought into actuality by an appropriate mental act of division.

Before moving to the examination of the spatially extended-which belongs to the phenomenal realm-and of its metaphysical correlate, we should consider two more aspects of space as an ideal entity. The first concerns the notion of a point as indicating a position in this framework, and the second the notion of distance in the abstract sense. Leibniz uses the idea of a referential system to capture the notion of position. In order to describe a position as indicated by a particular point A in space we need at least another fixed point B and a vector starting from B and ending at A. This already presupposes the notion of distance. Leibniz uses the notion in a way that involves both the idea of a vector and the idea of the vector's absolute value. He is not, of course, quite aware of that, but this should not worry us. He thinks of spatial distance as a relation between things spatially positioned; more generally, space is the manifold of those relations abstracted from their relata. A word of caution is necessary here. As we have already said, space as a manifold of spatial relations is not a simple aggregate of them, but an ideal entity that is prior to them. Spatial relations, though apparently constitutive of space, are posterior to and parasitic upon it. In a curious sense, although space is prior to spatial

[11] It seems that Leibniz considered as genuine continua only those that had an ideal status and were the subject matter of mathematical sciences. Consider, for example, the following passage from a letter of Leibniz to De Volder, dated January 19, 1706.

> ... a continuous quantity is something ideal which pertains to possibles, and to actualities in so far as they are possible. A continuum, that is, involves indeterminate parts, while in actuals there is nothing indefinite-indeed in them all divisions which are possible are actual ... Meanwhile, the knowledge of the continua, that is, of possibilities, contains eternal truths which are never violated by actual phenomena ... (L, 539; G, II, 282.)

relations, it is not, to Leibniz's mind, the absolute framework for measurement. Distance is a relation and is therefore dependent, even in the abstract, not upon the general idea of space, but upon its actual relata, considered as divested of their substantive peculiarities. In other words, what Leibniz wants to say is that a position in space is not fixed with respect to space as a whole, because there is not any particular referential system connected with it. A position in space is fixed only in relation to another position which *can be considered* as fixed. It is exactly this which makes Leibniz's theory of space not only a relational but a relativistic one; not, of course, in the full sense of the theory of special relativity. One of the basic postulates of the theory, however, is also a Leibnizian doctrine. According to Leibniz *there* is no *privileged spatial referential system.*

The priority of space as an ideal entity over spatial relations is not, for Leibniz, an excuse for admitting its absoluteness. Such a priority was forced upon him by worries about the composition of the continuum, and it does not necessarily imply the existence of a privileged referential system with respect to which absolute positions can be assigned to spatial entities.

In *Leibniz's Correspondence with Clarke* and in his "Fifth Paper" we find the following passage, in which Leibniz tries to define both place or position, and distance.

> 47. I will here show how men come to form to themselves the notion of space. They consider that many things exist at once, and they observe in them a certain order of coexistence, according to which the relation of one thing to another is more or less simple. This order is their situation or distance. When it happens that one of those coexistent things changes its relation to a multitude of others which do not change their relations among themselves and that another thing, newly come, acquires the same relation to the others as the former had, we then say it is come into the *place* of the former ... And supposing or feigning that among those coexistents there is a sufficient number of them which have undergone no change, then we may say that those which have such a relation to those fixed existents as others had to them before have now the *same place* which those others had. And that which comprehends all these places is called *space*. (A, 69; G, VII, 400.)

What makes the passage interesting is that Leibniz tries to define place or position by explaining what occupying the *same place* means. He does it by using the idea of the same referential system. A place cannot be defined in an absolute way with respect to a privileged referential system. On the contrary, it is defined with respect to a locally fixed, but relative system of reference. The set of objects which remain at fixed positions appear to do so because their mutual spatial relations remain unchanged for an appropriate period of time.

To put it metaphysically, the monads which are constitutive of them represent one another as spatially situated, in a way that remains the same for a corresponding appropriate period of monadic change. Furthermore, when we say that a certain something A is at the *same place* as a certain something B was before, we mean that A replaced B as the corresponding relatum in an otherwise unchangeable set of spatial relations, the other relata of which remained the same. If A and B were monads (as represented in space) we would say that B is now at the *same place* that A was before, if B represents the monads, which are constitutive of the fixed referential system, as spatially situated with respect to B itself, in exactly the same way as A was representing them before as spatially situated with respect to A itself. Similarly, the monads constitutive of the referential system would represent B as spatially situated the same way they were representing A as being situated. A place is fixed when a corresponding set of relative phenomenal spatial relations is fixed. And such a set of relations is fixed if the corresponding representations of the monads constitutive of the objects phenomenally placed this way are fixed accordingly. The distance of a monad A from a monad B, as they mutually represent each other, is described exactly by their phenomenal spatial relation. To put it more precisely, according to Leibniz, their distance is exactly their phenomenal spatial relation, which can be considered as reducible to relative positional predicates inherent in the corresponding monads. Such a relation, and therefore such a distance, is relative, because it depends upon the representational structures of the corresponding monads and not upon the existence of an absolute container (absolute space), a part of which separates them. The measurement of such a relative distance can only take place at the phenomenal level and it involves[12] the art of the comparison of the distance with a given unit of measurement. Concerning the realm of the ideal, it is enough to say that, for Leibniz, place, distance, and measurement as we know and use them

---

[12] Leibniz was right in thinking that the denial of the reality of space does not imply the denial of the possibility of spatial measurement. Consider, e.g., the following passage, from *The Leibniz-Clarke Correspondence,* Leibniz's "Fifth Paper" in which he insists that "relative things have their quantity as well as absolute ones."

> ... As for the objection that space and time are quantities, or rather things endowed with quantity, and that situation and order are not so, I answer that order also has its quantity; there is in it that which goes before and that which follows, there is distance or interval. Relative things have their quantity as well as absolute ones. For instance, ratios or proportions in mathematics have their quantity and are measured by logarithms and yet they are relations. And therefore, though time and space consist in relations, yet they have their quantity. (A, 75; G, VII, 404.)

at the level of phenomena have their ideal counterparts, as we know and use them in the framework of the science of space, i.e., Euclidean geometry.

The position of Leibniz concerning the nature and the structure of space-time can be characterized as *relational*. Space-time is not the absolute container of the movable, extended bodies of the Newtonian universe. "Space" for him "is the order of coexisting phenomena as time is the order of successive phenomena" (L,604; G,II,450). Time and space in the world of the well-founded phenomena "consist in relations" (A,75; G,VII, 404). A similar view was expressed by Huygens who also rejected the Newtonian position. The Newtonian view about the absoluteness of space-time was defended later by Euler and Kant[13]. Yet the arguments they used were different from those used by Newton. Berkeley, on the other hand, in the framework of his system, wherein a peculiar version of discreteness was proposed for space and time, rejected firmly the idea of an absolute space-time. In the nineteenth century the controversy between the two views continued. The dominant position, especially among the scientists of the time, was that space-time was the absolute container of things, objects and events. It was a well established view that Newtonian mechanics faithfully expressed the structural features of the world of physical objects and processes. Yet, nineteenth century was Maxwell's century. Although Maxwell, himself, was ambivalent with respect to Newton's notion of absolute motion, his scientific work seriously threatened the dominant view that space-time was absolute. Little by little the absolutists were loosing ground. Mach attacked them fiercely leaving no room for the idea of an absolute space-time. Conditions were right for, it seems, the final blow. The beginning of the twentieth century was marked by the appearance of the grand new theory, that is, the theory of relativity. Poincaré's ideas and Einstein's monumental work eventually dominated the scene. The absolutists were defeated at last and the ideas of Huygens and Leibniz appeared as finally winning. The phrase "appeared as finally winning" does not imply the existence of some kind of doubt about the defeat of the absolutists. On the contrary, their defeat was and is certain. By such a phrase we simply want to point out that Einstein's theory of relativity is, among other things, a firm negation of the Newtonian idea of space-time as an absolute container of objects and events with an absolute referential system fixed with it, rather than

[13] Although Kant argued for the absoluteness of space-time and so, in a sense, took a view opposite to that of Leibniz, it seems that Leibniz's influence on him (through Christian Wolff) was rather decisive. Such an influence is more particularly evident on Kant's consideration of space-time as the frame not of what *there is* but of what *can be phenomenally known*. Additionally, Leibniz's influence could also be observed in the way Kant viewed the relation between space and Euclidean geometry.

a vindication of the Leibnizian relational account of space-time and motion. According to John Earman (*World Enough and Space-Time*, p.7):

> In the early twentieth century Poincaré and Einstein set their authority against the absolutists, thus helping to fashion the myth that relativity theory vindicates the relational account of space, time and motion, favored by Huygens, Leibniz and their heirs.

It is, by now, generally accepted that in a strange way, *The Leibniz-Clarke Correspondence* was not only the result of the bitter struggle between Newton's absolutism and Leibniz's relationism concerning space-time and motion but also, the battleground where this same struggle developed. It was a "bitter struggle" not only because of what was at stake but also because of the priority and the plagiarism feud between the two great men, which preceded the correspondence, and concerned the discovery of the infinitesimal calculus. Relatively recent historical scholarship (Koyré and Cohen, [84], 1962) has uncovered and provided evidence to confirm the long-held suspicion that Newton actually wrote Clarke's responses to Leibniz. There are people, of course, like Hall (see [66],1980) who think otherwise. In any case it is fair to say that Newton was directly involved either in writing them himself or closely supervising their writing. So Clarke's response to Leibniz had one way or another Newton's complete approval. What is then going on in *The Leibniz-Clarke Correspondence*, what was scientifically at stake in this controversy? The answer is not difficult. It was the appropriate conceptualization of the nature and the structural properties of space-time.

Which were the main characteristics of Newton's conception of absolute space and time? Earman gives a quite convincing answer to this question. According to him the main characteristics of Newton's conception of absolute space and time (see [31] p.7), are the following four:

Absolute Motion: space and time are endowed with various structures rich enough to support an absolute , or non-relational, conception of motion.

Substantivalism: these structures inhere in a substratum of space or space-time points.

Nonconventionalism: these structures are intrinsic to space and time.

Immutability: these structures are fixed and immutable.

The most important of the above four interrelated but distinct characteristics are the first two. Connected with the first one is the Newtonian claim that there is an absolute reference frame which is responsible for the uniqueness of the identification of spatial locations through time. According to Earman, "as a result there is an absolute or well-defined measure of the velocity of individual particles and a well-defined measure of spatial separation for any pair of events" (Ibid., p. 11). The theme of *substantivalism* in the Newtonian writings is emerging through the claim that space-time is a substance. Such a claim should be cashed out in terms of what is for space-time to be a substance. Space-time is a substance in the sense that it forms a substratum of space-time points and regions with spatio-temporal relations inherent in it and it is the case that similar relations among physical events and processes are parasitic upon the first ones. The characteristic of *nonconventionalism* is the one according to which the space-time structures are intrinsic to it. Thus things like absolute simultaneity and absolute duration, as space-time structures, are important parts of the intrinsic spatio-temporal basis of the actual world of physical objects and of physical events and processes. The fourth characteristic, *immutability* that is, is the one according to which the space-time structures are the same from time to time in this world and from a physically possible world, this actual world of ours included, to any other physically possible world.

Leibniz's position concerning space-time in the world of phenomena constitutes a firm denial of the Newtonian position. There is no absolute container of things, events and processes. The world of phenomena is well-based upon the real world of monads as representing one another as representing. What appears to be the spatio-temporal framework of the world of phenomena is nothing over and above the set or the manifold of the phenomenal spatio-temporal relations which are direct among bodies and events. Such phenomenal relations can be analyzed in terms of, and reduced to, monadic representational predicates inherent in the monads the way we described in Chapter I, Section B. Yet they are not and cannot be parasitic on pre-existing relations. Such relations do not exist because there is not any spatio-temporal substratum to which they inhere. Leibniz's relationism, as it is expressed in his mature writings and especially in *The Leibniz-Clarke Correspondence,* is based upon two basic contentions. First, all motion of bodies is relative and all spatio-temporal relations are direct among physical objects and events. Secondly, space-time is not a substance, that is, there is not a substratum of space, time and space-time points, endowed with spatio-temporal relations ontologically prior to spatio-temporal relations among physical bodies and events.

As we have already said, Leibniz considers space to be an ideal entity. It belongs to the set of innate ideas which is the subject of the science of mathematics. We come to form the notion of space, Leibniz holds, out of what we consider to be spatial relations abstracted from their relata. Yet, from the epistemological point of view, space is neither a mental construct, nor an immediate product of abstraction. It is an innate idea of which we are helped to become aware by our everyday experience of the mutual simultaneous arrangement of extended objects. It does not correspond to a unitary real entity and it is the idea of the simultaneous order of coexistence of things as they are represented by us.

The spatially extended, on the other hand, belongs to the realm of the phenomena. When we perceive an extended object-and here we predominantly mean visual perception-we simply represent an infinite multitude of monads as if they were positioned relatively to one another, so that we get the impression of a more or less continuously extended and undifferentiated whole. Yet *monads are not positioned in space*, they merely *represent* one another as so positioned. Such a representation has two aspects, the one phenomenal, the other real. A representation by a monad A of the world of the other representing monads (itself included) is a viewing of this same world as having spatio-temporal features. The world of monads as representing one another is represented *by us* so that it appears *to us* as the spatio-temporal world of well-founded phenomena. On the other hand, a representation by a monad A of the real world of representing monads is a *real* state of the monad A, which together with the other coexisting[14] monads, being in appropriately corresponding states, forms a real state of the world. This real state of the world is represented by every monad perspectively; that is, it is represented by every monad as if each one, as a representor or viewer, were at the center of viewing the *same* world of phenomena. The perspective centers corresponding to any two such *simultaneous* viewings by, e.g., monads A and B, are of course different. They differ because, first, the monads A and B have different representational structures and second, because at the level of the phenomena, each one of these monads sees the world from a unique perspective.

Before we move on we should discuss in more detail spatial perspective representation. As we said, a monad A views the world of phenomena as if it itself were at the center of such a viewing. The abstract model that Leibniz presumably had in mind was that of a point from which all possible straight lines are drawn, so that space, as a three-dimensional Euclidean continuum, is filled with them. It is in this sense, we think, that Leibniz calls the point of view of a monad a *mathematical point*. Every such line could be called a *perspective line* and the monads represented as positioned on it *perspectively linearly equivalent*. We can think of at least two possible ways of metaphysically interpreting perspective representation. One would be that of

[14] Where coexistence means simultaneity at the level of monadic change.

deploying the machinery of the *degrees of distinctness* of representations and the other, that of using the *positional predicates* inherent in the monads. Both can be accommodated in Leibniz's system: (a) A perspective line could be thought of as the phenomenal counterpart of a series of indirect representations which as a whole and under the appropriate circumstances is characterized by a *special degree of distinctness*. By "appropriate circumstances" we mean those which result to a representation of a series of monads by any monad, which phenomenally belongs to such a line and "looks" at the world in such a way, that this same line appears to be at the center of the monad's visual field. (b) A perspective line could be thought of as the overall resultant of a series of indirect representations in the framework of one and the same representing, so that each such indirect representation is of a monad as specifically spatially positioned on such a line. That is, the *positional predicates*[15] *in its monad,* which phenomenally specify such a line, *are* so to speak, *responsible for the appearance of linearity*. More specifically, such predicates play the role of the metaphysical counterparts of what we would theoretically call positional co-linear vectors.

From God's point of view there corresponds to the real world of monads representing one another a well-founded phenomenal counterpart. This counterpart is well-founded for two reasons. First, the correspondence is such that rules which govern it remain always the same[16]. Second, monads share the view of the *same* world of phenomena as a representation of the real world of representing monads, in a way that is perspective. Monads belong to the real world. Each one represents all the others from its point of view. That is, it is not only the case that a monad represents itself as spatially positioned, but it represents all the others so positioned, with itself at the center of the viewing. A momentary perspective representation of the real world of representing monads by a monad A, for example, is the internal viewing by this same monad of a particular representational state that it is in. The reality of all monads being in representational states corresponding to the one that monad A is in, is represented by A so that its internal viewing is that of an orderly three-dimensional arrangement at the center of which A is represented as situated. All these perspective representations conform to one another in such a way that any set of simultaneous representations by the existing monads corresponds to

[15] Via which each such monad represents all the others as spatially positioned with respect to itself.

[16] It is our view that such a correspondence has the characteristics of an isomorphism. In sections III.B and III.D we will give specific examples of what we mean by this in the cases of spatial and temporal density.

a phenomenal, continuous, three-dimensional arrangement of the monads, with no indiscernibles in it. The perspective representation of the real world of monads by each one of them is the perspective viewing of one and the same world of phenomena, a non-perspective viewing of which is God's privilege.

This description is perhaps too general; it does not give us enough information about the structure of the correspondence between the levels of the phenomenal and of the real, as related to the case of the spatially extended. In the next section we will construct a model with which we will attempt to throw more light on the nature of such a correspondence. More specifically, we will see how we can make sense of that which we will call the relation of *spatially between* as one that is based upon the representational reality of monads. Through it we will specify, in terms of the reality of monads as they harmoniously represent one another, what *density* means in the case of spatial extension. For the moment, we will continue the examination of both the levels of the phenomenal and the real, with their interweaving in spatial extension as our central theme.

As noted earlier, the spatially extended belongs to the realm of phenomena. Phenomena which are not the result of hallucinations, nor the subject matter of dreams, nor mere fantasies are, for Leibniz, well-founded. There are two characteristics that such phenomena possess. The first is of an epistemological, the second of a metaphysical nature.

(1) In his short paper *On The Method of distinguishing Real from Imaginary phenomena*, Leibniz offers a set of criteria for distinguishing real from imaginary phenomena. These criteria can be used for each phenomenon in question and involve the presence or absence of certain characteristics that the phenomenon should have by itself or in relation to those preceding or following it. Leibniz gives a protreptic description of these criteria:

> Let us now see by what criteria we may know which phenomena are real. We may judge this both from the phenomenon itself and from the phenomena which are antecedent and consequent to it as well. We conclude it from the phenomenon itself if it is vivid, complex, and internally coherent. It will be vivid if its qualities, such as light, color and warmth appear intense enough. It will be complex if these qualities are varied and support us in undertaking many experiments and new observations ... A phenomenon will be coherent when it consists of many phenomena, for which a reason can be given either within themselves or by some sufficiently simple hypothesis common to them; next, it is coherent if it conforms to the customary nature of other phenomena which have repeatedly occurred to us, so its parts have the same position, order and outcome in relation to the phenomenon which similar phenomena have had...

> But this criterion can be referred back to another general class of tests drawn from preceding phenomena. The present phenomenon must be coherent with these if, namely, it preserves the same consistency or if a reason can be supplied for it from preceding phenomena or if all together are coherent with the same hypothesis, as if with a common cause. But certainly a most valid criterion is a consensus with the whole sequence of life, especially if many others affirm the same thing to be coherent with their phenomena also. Yet the most powerful criterion of the reality of phenomena, sufficient even by itself, is success in predicting future phenomena from past and present ones, whether that prediction is based upon a reason, upon a hypothesis that was previously successful, or upon the customary consistency of things as observed previously. (L, 363-364; G, VII, 319-320.)

These criteria, especially that of internal coherence of a phenomenon and of coherence of the whole set of real phenomena, are impressively modern. If we carefully examine them, we find all the important ingredients that modern philosophical systems, with their distaste for heavily loaded metaphysics, cherish. Ideas such as those of inductive generalization, lawlikeness, predictability, intersubjectivity (as a substitute for objectivity) in the sense of a consensus of the majority of observers, etc., are all there. Yet, real phenomena cannot be called well-founded just because they happen to satisfy such criteria. Coherence of the real phenomena, Leibniz says, although a good alternative for truth so far as everyday practice is concerned, is not the most we can ask for[17]. A good explanation is not one that simply describes the difference between dreams and reality in terms of internal epistemological criteria such as coherence. A good explanation requires good metaphysics. And a good metaphysics, according to Leibniz, is our ultimate explanatory source. It is, indeed, the fountain of truth.

(2) The class of real phenomena is the same as the class of well-founded phenomena. Yet, though real phenomena can be distinguished from the imaginary ones through a set of epistemological criteria, they are well-founded not for epistemological but for metaphysical reasons. It is metaphysics that can explain the existence of such criteria at the epistemological level. Leibniz's theory of truth is representational. The coherence of the real phenomena is a by-product of their well-foundedness, and therefore, although it can provide us

---

[17] From the same short paper of Leibniz comes the following interesting passage:

> We must admit it to be true that the criteria for real phenomena thus far offered, even when taken together are not demonstrative, even though they have the greatest probability; or to speak popularly, that they provide a moral certainty but do not establish a metaphysical certainty (L, 364; G, VII, 320.)

with a set of practical criteria for recognizing what we would call reality, it is not a good substitute for a theory of truth. A real phenomenon is well-founded because it corresponds to a metaphysical reality, upon which it is grounded. And such a metaphysical reality is, for Leibniz, monads representing one another in a specific manner.

Let us consider again the spatially extended. More specifically, let us examine a particular extended object, the chair in this room, say. It is positioned in the middle of the room and will continue to be so positioned until, e.g., someone decides to move it. What does it mean to say that the chair is so positioned? It means simply that the chair does not change its distance from fixed characteristics of the room, such as its walls, floor, and ceiling. Additionally, the chair retains its shape. But these features are internal to the phenomenon of the chair being positioned in the middle of the room. The question why such a phenomenon is well-founded cannot be answered in the framework of a language, the semantics of which is dictated by the phenomena themselves. The Leibnizian answer to this question is that the chair is not a real entity, but in some sense consists of real entities. To the phenomenon of the chair positioned in the middle of this room corresponds the reality of an infinite multitude of monads, represented by us (as representing one another) as if they were in fixed spatial relations among themselves, and in fixed spatial relations to the monads constitutive of the walls, the floor, and the ceiling of this room.

These monads are represented by us as unextended, but we do not perceive them distinctly enough, i.e., we cannot be aware of them separately. We perceive them confusedly and *en masse*, as extended wholes. That is, we have a distinct representation of the whole, but not of each monad separately. It is an empirical fact that sense perception and, more specifically, visual perception is that of extended objects. There are three reasons for this[18], which together can answer the puzzle posed by such a situation. First, every monad, according to Leibniz, has, associated with it, a *minimum perceptibile*[19], a boundary, so to speak, beyond which we cannot distinguish finer details in a spatially extended object. It is a defect of the representational structure of every monad, although the minimum perceptibile can differ from species to species, or monad to monad. It is true that Leibniz does not refer to such a minimum explicitly, but there are passages where such a thing is tacitly implied as characterizing the ability of any created being to perceive the world of phenomena[20].

---

[18] The first two of which we discussed in chapter I.

[19] Here we mean, in particular, a *minimum perceptibile* connected with visual perception.

[20] See, e.g., L, 536; G, II, 268.

Additionally, he does not offer any metaphysical rationale for it. It seems that he simply considered it as one of the many imperfections created beings are endowed with. Second, monads, as represented, form a spatially extended plenum. No matter how small a spatially extended object or area, an infinity of monads are represented in it as positioned continuously. A monad, in other words, cannot be viewed by another monad as isolated from its densely populated enviroment. Monads are represented so that consecutive parts of what appears as extended are infinitesimally close. That is, there is no finite distance separating consecutive parts in the physical plenum which would correspond to a, so to speak, representational vacuum. Third, it seems that according to Leibniz, and as a byproduct of the first two reasons, contrasts due to the different degrees of the distinctness or confusedness of what is represented cannot be observed as existing between a monad and all the others as represented. Such contrasts can only be observed between appropriate multitudes of monads which correspond to what appear to be separate extended objects. For instance, we are able to perceive distinctly the chair in this room without being able to discern our separate representations of its constitutive monads. According to Leibniz, this is due to our inability to represent distinctly an infinitude of items as differing from one another. One of the differences between us and God is precisely this; he can and we cannot reach the infinite. God has, as it were, a definitional knowledge of all the infinitely many items that constitute the real (or the possible), a knowledge which humans, or more generally created beings do not have and cannot hope to attain.

Real phenomena, again, are well-founded because they are grounded upon the metaphysical reality of monads representing one another. A characteristic of this metaphysical reality, and therefore a feature of the explanatory schema for the well-foundedness of real phenomena, is that the representational structures of all monads are so well coordinated that their representations of one another as representing, though perspective, conform to one another harmoniously. If, for example, two persons in normal circumstances were in the *same* room, they would both see from different perspectives the *same* chair as situated in the middle of the room. Metaphysically speaking, their representational structures would contain isomorphic features concerning their representations of monads constitutive of the chair, the walls, the floor, and the ceiling of the room, as well as of what appears to be their spatial relations.

Leibniz sees spatial relations as basically relations of distance. But relations of distance (and for that matter relations in general) do not really exist between monads. They are appearances, parts of the well-founded phenomena of extendedness, grounded upon the reality of monads as representing one another. So the viewing of the phenomenal spatial relations we mentioned above by each of the two people in the room would be simply the result of appropriate representational interweaving of the monads constitutive of the chair, the walls, the ceiling, and the floor of the room representing one another as spatially situated. Each of the two people would be a representor, internally viewing his own representing of what really or metaphysically obtains. Their representations would be isomorphic and perspective. Isomorphic, because from God's point of view they can be isomorphically mapped onto one another and perspective, because they would contain different representational features, which can be described as corresponding to the phenomenally different positions that the two persons would occupy in the phenomenal spatial region of the room.

Leibniz scholars in general, under Russell's influence, seem to be confused about continuity as extended to the level of the real. The ultimate reality, after all, is that of a *discrete* multitude of monads which are not spatially situated. Thus, they recognize continuity of space and, of course, continuity of the spatially extended as definite Leibnizian doctrines, without paying much attention to whether or not the monadic reality is continuous in an analogous way. They appear to be content with admitting that Leibniz holds that monads represent perspectively all the others so that no spatial vacuum exists at the level of the phenomena, and do not pursue the matter any further. They are aware, of course, of the Leibnizian contention that the non-existence of a vacuum at the level of the phenomena indicates the non-existence of a vacuum at the level of the real. It seems, on the other hand, that they hesitate to admit that monads, though not existing in space, could have been positioned in what we could call *sprace*, an entity with a structure isomorphic to that of space. And they are right on this point. There is nothing existing by itself which would serve as the metaphysical container of monads and which could be represented by them as space. On the other hand, they are wrong in thinking of the totality of monads as nothing but a discrete whole, which does not form at the level of the real a continuum parallel to the continuum of the phenomenally extended. Consider Russell, for example; he thinks that continuity is in one sense denied by Leibniz[21]:

> In spite of the law of continuity, Leibniz's philosophy may be described as a complete denial of the continuous. Repetition is *discrete*, he says, where

[21] Parallel views, concerning what is taken as the denial by Leibniz of the continuity of the world of phenomena as based upon the discreteness of the real world of monads, one can find, for instance, in [68].

> aggregate parts are discerned, as in number: it is continuous where the parts are indeterminate, and can be assumed in an infinite number of ways (N.E. p. 700; G.IV. 394). That anything actual is continuous in this, Leibniz denies; for though what is actual may have an infinite number of parts, these parts are not indeterminate or arbitrary, but perfectly definite(G.II. 379).Only space and time are continuous in Leibniz's sense, and these are purely ideal. In actuals, he says, the simple is prior to the aggregate; in ideals the whole is prior to the part (G.II.379). Again he says that the continuum is ideal, because it involves indeterminate parts, whereas in the actual everything is determinate. The labyrinth of the continuum, he continues-and this is one of his favorite remarks-comes from looking for actual parts in the order of possibles, and indeterminate parts in the aggregate of actuals.([128], p.111) .

It is a fact that there exist passages in Leibniz's writings which give the impression that continuity is indeed in one sense denied by him. Let us consider, for instance, the following passage from a letter of Leibniz to De Volder, dated October 11, 1705:

> Matter is not continuous but discrete, and actually infinitely divided, though no assignable part of space is without matter. But space, like time, is something not substantial, but ideal, and consists in possibilities, or in an order of coexistence that is in some way possible. And thus there are no divisions in it but such as are made by the mind, and the part is posterior to the whole. In real things, on the contrary, units are prior to the multitude, and multitudes exist only through units. (The same holds for changes, which are not truly continuous). (R, 245; G, Il, 278-279.)

Or let us consider the next passage from a letter of Leibniz to Des Bosses, dated July 31, 1709, where he seems wanting to make a sharp contrast between *continuity* of space and *discreteness* of mass.

> Space, just like time, is a certain order ... which embraces not only actuals, but possibles also. Hence it is something indefinite, like every continuum whose parts are not actual, but can be taken arbitrarily, like the parts of unity, or fractions... Space is something continuous but ideal, mass is discrete, namely, an actual multitude, or being by aggregation, but composed of an infinite number of units. In actuals, single terms are prior to aggregates, in ideals the whole is prior to the part. The neglect of this consideration has brought forth the labyrinth of the continuum. (R, 245; G, II, 379.)

In these passages Leibniz is thinking of the ideal mathematical nth-dimensional continua, where n is equal to 1 or 2 or 3, as the only genuine ones. It is true that such continua were thought by him to be genuine because they possess the characteristic af *uninterruptedness*. In the passages quoted, Leibniz goes even further by equating continuity with such uninterruptedness. It seems that he thought that in running through a continuum we should not stumble upon existent discrete entities. But this is misleading. Continuity, for Leibniz, means primarily density, and this is definitely a property that both the phenomenal and the real levels of his metaphysics possessed. Although Leibniz favored the ideal continua over the phenomenal or real ones-even reserving, most of the time, the term "continuum" for the former, as signifying uninterruptedness-he did not really deny continuity as a governing principle either to the real or to the phenomenal. By insisting, for instance, that monads are graded continuously, according to the relative confusedness or distinctness of their representational structures, he was admitting that we can have a real continuum, *which* is *discretely composed.* It is discretely composed in the sense that a complete decomposition[22] of it would result in an infinity of point-entities, which would already exist before such a decomposition took place. In such a continuum, all the divisions that are possible, are actual and density means that in between any two *actual* point entities there is always a third *actual* one.

Returning to our initial contention, we think that Russell was wrong when he insisted that Leibniz denies continuity in the case of the actual, where "actual" signifies, in some confused way, both the real and the phenomenal. What Leibniz wanted to emphasize in the case of actual continua was their lack of uninterruptedness, which he quite misleadingly sometimes equates with continuity. We think Russell became confused because Leibniz, in trying to solve the problem of the composition of the continuum, reserved the term "continuum" for the ideal ones only. Let us consider the following short passage from a letter of Leibniz to De Volder, dated January 19, 1706.

> ... the knowledge of the continua, i.e., of possibilities, contains eternal truths, which are never violated by actual phenomena ... (L, 539; G, II, 282).

The passage is illuminating in another respect also. It shows that Leibniz believed that continuity is never violated by well-founded phenomena. That is, continuity, not as uninterruptedness, but as infinite divisibility or density governs the realm of well-founded phenomena also. Furthermore, well-founded. phenomena are grounded upon the reality of monads representing one another. It would therefore be unwarranted to say that Leibniz would deny continuity as governing the realm of the real either. If well-founded phenomena are to conform to the principle of continuity, the monadic reality has to conform to it too. It would be at least paradoxical to say that although

---

[22] A complete decomposition of it can be thought of as an act of collecting together all its constitutive points by ignoring or destroying all their mutual positional relations.

Leibniz thinks of the phenomena as continuously structured, he also thinks of them as representations of a discrete and discontinuous reality.

Monads in fact exist, they are always in particular representational states and these are always on the verge of changing. The mutual agreement of these states is a fact for Leibniz, based upon God's decision to create the best of all possible worlds, a world characterized by, *inter alia*, a pre-established harmonious synchronic and diachronic coordination of the representational states of all created monads. Each monad should be thought of not on the model of an empty shell, but as a unique substantial unit characterized by a specific representational structure. Such a structure should be thought of as the enfolding equivalent of a blueprint, or of a life plan which characterizes, in a non-temporal sense, the essence of the monad.

A momentary state of the real world of monads can be defined as constituted of all the monads at particular representational states, so coordinated that all of them taken together correspond to what appears to be the world of phenomena at a particular phenomenal time moment. A particular momentary state of a monad is a static perspective representing of all the other monads as perspectively representing one another at the same phenomenal time moment. Such a manifold of mutual synchronic perspective representations is seen by God non-perspectively and non-temporally. Perspective representation is a characteristic of the created monads and has two aspects. First, it is a reality in itself in the sense that a real momentary state of a monad can be described as that of a perspective representation. Second, a perspective representation is an internal perspective viewing of the world as represented by a monad. The monad represents all the others (itself included) as if it were positioned at a particular spatial point from which it views them as analogously positioned and filling all possible point-positions in what appears to be a three-dimensional Euclidean continuum.

What appears to us as spatially extended is a phenomenon *bene fundatum*. As we noted, both at the level of phenomena and at the level of the real there is a continuous plenum. The world we experience as extended does not contain vacua. *The real world* of *monads does not contain vacua either*. But this can be expressed in a variety of ways. There is a special way to phrase this contention, as related to the continuity of the spatially extended. What we want to say is that the world of monads, although not spatial, has a structure isomorphic to the structure of the spatially extended. That is, there are no vacua in this monadic structure in a way parallel to the absence of vacua in the world of the spatially extended.

It is wrong to think of the reality of monads on the model of a discrete, discontinuous, and unstructured infinite whole. If we were to think of them this way, then the contention that the world of monads is a plenum would either be meaningless or, if we had to give it some meaning, parasitic upon the thesis that the phenomenal world of the spatially extended is continuous. That is, we would simply commit ourselves to the following weak thesis: the world of monads does not contain vacua of forms (as this is related to the phenomenally extended), because at any particular phenomenal time moment and to every possible point position in a three-dimensional Euclidean continuum, corresponds exactly one monad phenomenally occupying that position at this moment, and *vice versa*. But monads are not simply empty shells. They are characterized by their particular representational structures. A certain part of that structure is responsible for the spatial perspective representation which characterizes each one of them.

We can think of the totality of monads and picture it on the model that modern set theory provides us with. What is a Euclidean three-dimensional continuum according to set theory? It is an infinite set of points[23] A, together with an infinite set of particular mutual relations R. Such a set-theoretic ordered pair (A,R) can be pictured by the classical Euclidean three-dimensional continuum E, so that a *Cartesian* one to one correspondence can be established between (A,R) and E. To each actual point of A would correspond exactly one possible point-position in E and to each possible point-position in E exactly one actual point of A, so that the set of actual relations of the points of A would be *isomorphically* mapped onto the set of all possible relations of the possible point-positions in E. E can be thought of as prior to such a correspondence and therefore prior to its partition.

The totality of monads considered as devoid of any representational structure can be thought of on the model of the unstructured set A, i.e., on the model of A stripped of its relational structure as described by the set R. But monads do have a representational structure. They always represent one another as representing. Let us consider their world at a particular momentary state. They represent one another as representing in a spatially perspective way. The set of all of these perspective representations plays, at the level of the real, a role analogous to the one the set of positional relations R plays in the case of the three-dimensional set-theoretic continuum. The only difference is that in the case of the world of monads we do not have spatial or positional relations. But this should not alarm us. The metaphysical substitute for spatial

[23] Details about its cardinality are not relevant to our discussion. It is enough to say that the cardinality of such a set is equal to the cardinality of the real line.

relations is the perspective representations as real states of the monads. After all, relations can be reduced to monadic representations the way we described in chapter I. We will have the opportunity to consider a specific example of how such a thing can be done in the next section.

It is in that sense, we think, that one should insist that monads at the real level form a continuum isomorphic to the phenomenal continuum of the spatially extended. It is a continuum structured discreetly. The spatially extended is also a continuum, but it is structured discretely in a parasitic way. That is, when Leibniz insists that the spatially extended is "actually infinitely divided" he confusedly runs together two isomorphic, but metaphysically different levels, namely, the level of the real and the level of the phenomenal. The spatially extended is a phenomenon *bene fundatum*. As a phenomenon it is a representation, confused in its minute details, of an infinite multitude of monads in particular representational states. Every monad in that infinite multitude is represented in the framework of such a total representation. The spatially extended is "actually infinitely divided," in the sense that such a phenomenon consists of all the representations of the corresponding monads in such a way that "in between" any two monads, as represented, there exists a third one, as represented. Such density at the phenomenal level corresponds to density at the real level, in that representations are not of monads as empty shells, but of monads in particular representational states. Our real representational states are in one sense responsible for the way we are represented by others as spatially positioned. In other words, a monad A represents another monad B as truly occupying a particular spatial position because it represents B as being in a particular spatially perspective representational state, while B is really in this state. To the phenomenal density corresponds real density, and to a three-dimensional continuous phenomenal world corresponds a three-dimensional real world consisting of all the monads at particular, spatially perspective, simultaneous states. At the level of the real density means that for any two monads at particular simultaneous, spatially perspective representational states there exists a third monad in a particular spatially perspective representational state so that at the phenomenal level the third is represented as positioned "in between" the other two.

As it has been said, in the next section we will construct a model which makes clearer the interpretation of Leibniz we propose. More specifically, we will attempt to give an account of what the correspondence between the level of the real and of the phenomenal amounts to in the case of the relation "spatially between." We will make use of the notions of distinct, confused, and indirect representations which were discussed in chapter I. Finally, we will give a formalization of the interconnection of the two levels in the case of what we consider to be a particularly important manifestation of continuity, viz., density of the spatially extended.

B. THE PHENOMENAL RELATION OF "SPATIALLY BETWEEN", ITS METAPHYSICAL BASIS, AND DENSITY AS A PROPERTY OF THE SPATIALLY EXTENDED

What makes real phenomena well-founded is their correspondence to a non-spatio-temporal metaphysical reality, formed out of an infinite multitude of indivisible representors being at specific representational states and striving toward future ones. In order to make sense of such a correspondence, as it affects the spatially extended, we will examine how the phenomenal relation of "spatially between" can be based upon representational facts belonging to the metaphysical realm. According to Leibniz, relations are phenomenal bridges between phenomenal entities. Monads are not related through a set of metaphysical entities existing over and above them. Each monad is windowless and self-sufficient. Yet , we can have a metaphysical surrogate for relations, namely, synchronic and diachronic configurations of representational states of the monads which, when represented, give rise to the phenomenal reality of relations. In other words, one represents as relations representational facts of the form "S_1 represents (in unique *manner* M_i) S_2". Therefore, relations are not *real* but phenomena *bene fundata*.

Let us briefly return to Sellars' analysis of the Leibnizian relations as adopted and expanded in chapter I. Let us consider again the relational statement S_1 is R_i to S_2 where S_1 and S_2 signify particular monads. According to Leibniz, for this relational statement to be true, there must be an R_i -to-S_2 representationally inherent in S_1. We can once again summarize the Sellarsian interpretation of Leibniz's contention that relations are reducible to representational facts of the above sort[24].

(a) R_i -to-S_2 is inherent in S_1 in the sense that it is a representing by S_1 of some particular aspect of the world of monads as representing. That aspect is determined only by *what there is*, namely, by the world of monads as they represent one another as representing.

(b) S_2 is inherent in S_1 as a representing which is a "part" of the representing R_i -to-S_2. Using Cartesian terminology, S_2 is inherent in S_1 as an

---

[24] This is not to say that all relations are dyadic. Polyadic relations can be analyzed in a similar way.

objective or *representative* being in the middle of a more complex representing which is the R_i -to-S_2. It corresponds to the *formal* reality of the monad S_2 and belongs to the more complex representing R_i -to-S_2 which corresponds in its turn, by way of being their representation, to particular representational facts of the world of monads, namely, to particular complexes of concurrent representational states of the inhabitants of the same world.

(c) Relational truths rest on facts of a representational nature. Such facts are predicative and not relational.

(d) To the *prima facie* real relation R_i corresponds a specific manner of representation M_i which is defined by structural characteristics of a representational nature. Such characteristics are directness or indirectness, as well as relative distinctness or confusedness, of the representation as a whole, or with respect to its parts.

In order to see how particular monads perceive extended bodies we have to examine particular examples of important relations connected with extension. One such example is the relation of "spatially between." It is important to notice that this phenomenal relation constitutes the backbone of the Leibnizian contention that the extended phenomenal reality is not only infinitely divisible, but actually infinitely divided. This contention is equivalent to the thesis that the world of the phenomena is densely populated. A densely populated world, after all, is precisely where the relation of "spatially between" is literally at home.

All relations, including spatial ones, are, according to Leibniz, phenomenal, i.e., they are well-founded appearances. They are based upon representational facts which in their turn are special ways of representing the monads as representing one another. It is precisely this aspect of the representational facts that requires their examination via the notion of indirect representation introduced in section I.C. The question now arises whether Leibniz describes a way of associating the phenomenal with the real, so that we can determine the nature of the correspondence between the reality of the representational states of the monads and the phenomenal world. Nowhere in his writings does Leibniz specify the particulars of such correspondence. He seems content with giving us only some very general features of a strategy for finding its nature. Such features are based on the notion of representation as involving a perspective isomorphism between what is represented, by different representors, and the representings themselves. Additionally, by insisting that the only reality is that of the monads representing one another, he offers us, without himself fully realizing it, the notion of indirect representation as one of the basic tools for discovering the nature of the correspondence between the real and the phenomenal. Finally, such features of the representations as confusedness, distinctness, minuteness and insensibility, help us explain differences among perceptions of the same reality. These differences are due to the individual discrepancies of the representational structures of the individual representors themselves. The fact that Leibniz nowhere in his writings specifies a complete set of particular characteristics of the correspondence between the real and the phenomenal is something we must simply accept. His silence on this matter leaves us with the task of constructing a plausible model for such a correspondence in a way faithful to his philosophy. The construction of such a general model is very difficult, perhaps impossible. We will therefore restrict ourselves to the construction of a model for the special case of the important relation "spatially between," a task which we think can be carried out.

Let us consider three monads S_1, S_2, S_3 which are perspectively linearly equivalent and are represented as positioned in the same perspective line. Let us assume that I perceive S_2 as *spatially between* S_1 and S_3. The corresponding real counterpart of such a relation could be described as consisting of the following three representational facts.

1. S_1 represents S_2 as representing S_3.
2. S_3 represents S_2 as representing S_1.
3. S_2 represents both S_1 as representing S_2 as representing S_3, and S_3 as representing S_2 as representing S_1.

For reasons of convenience we can refer to this situation by saying that S_2 is *representationally between* S_1 and S_3. It is now obvious that the real counterpart of the phenomenal relation "S_2 is spatially between S_1 and S_3" could be the sum total of the above three representational facts, in other words, the composite representational fact described by the expression "S_2 is representationally between S_1 and S_3." When I correctly (truly) perceive S_2 to be spatially between S_1 and S_3, I indirectly represent S_2 as being representationally between S_1 and S_3, exactly when "S_2 being representationally between S_1 and S_3" obtains in the real world of monads representing one another.

A new question now arises, namely, how we know that this counterpart of the phenomenal relation we constructed ("S_2 is spatially between S_1 and S_3") is the right candidate for being the relation's metaphysical (i.e., real) counterpart. The answer is that we do not know, but only that it is a correct one and that we could possibly come up with other, equally correct, suggestions. Yet the fortunate thing is that after some thought we can convince ourselves that every other correct model for a real counterpart is isomorphic to, in a weak but quite a fundamental way, the one given above; that is, every other model for a real counterpart has to preserve the symmetry with respect to the central monad S_2, and this is exactly what matters. Such a symmetry is what makes impossible any conceivable permutation of the relative positions of the three perceived monads. In other words, all possibly correct models form an equivalence class out of which we have decided to pick a certain candidate that we agree to consider as the *right* one.

At this point another objection could be raised. One could doubt the validity of the above proposal on the grounds that perceivers do not distinguish monads in their perceptions[25]. Such an objection could be met by pointing out that although it is true that we do not distinguish monads in our perceptions, we nonetheless have confused perceptions of them, which means that we represent them one way or another. We represent them by representing the complexes they form. The fact that we cannot distinguish the monads constitutive of a billiard ball does not mean that these monads are not represented as being there. After all, perceiving a billiard ball B_2 as spatially between two other balls, B_1 and B_3, which are equal to it, is a result of confusedly perceiving every monad of B_2 as spatially between two other monads, one belonging to B_1 the other to B_3. The expression "the monad belonging to billiard ball B_i" is of course shorthand for the expression "the monad belonging to the metaphysical reality represented as the billiard ball B_i."

The proposed model for the metaphysical correlate of the phenomenal relation "spatially between" is correct for one more reason. In its construction we took into account the fact that a monad A always represents another monad B through the eyes, so to speak, of intervening ones, i.e., through ones which are represented as being spatially in between A and B. So the monad S_2 cannot represent S_1 as representing S_3, nor S_3 as representing S_1 without S_2 itself being represented as intervening in the series of the corresponding indirect representations. So S_2 represents, in a symmetrical fashion, both S_1 as representing S_2 as representing S_3, and S_3 as representing S_2 as representing S_1.

Because of the principle of continuity, an infinitude of other monads is represented as being spatially in between S_1 and S_3, or, for that matter, as being spatially in between any two monads. That is, any indirect representation is necessarily of infinite length. We can ignore the complications arising from such a truth by considering only finite sequence segments of any such indirect

---

25 We mean here sense perceptions and, in particular, visual perceptions.

representation. Such an option is indeed available to us. It is, after all, correct to say that a monadic representor can only distinctly represent finite sequence-segments of such an indirect representation. Additionally, there is no theoretical reason for not using such finite sequence-segments for the purpose of constructing appropriate representational models, without ignoring the fact that they are genuine parts of indirect representations, which not only have infinite length but also continuous structure. The fact that a monad A represents a monad B through the eyes of all the monads, which are represented as spatially intervening between them, does not run against the view that A represents B indirectly through the eyes of any specific monad C which belongs to the metaphysical reality represented as spatially in between A and B. Finally, that a monad A indirectly represents a monad B by representing the phenomenally intermediate ones as representing it, it is not something to which Leibniz would object, although he does not explicitly express such a view. Yet, this is surely what he means when he insists that the distinctness of a representation fades out as the phenomenal spatio-temporal distance increases and that the soul represents all the other monads indirectly, by representing the ones which are representationally nearer to it as representing them. We think that the unconsciousness of the representations of the very distant spatial objects that we have, is, according to Leibniz, due to the fact that the greater the number of monads represented as intervening between us and them (through the eyes of which, so to speak, we come to see them), the fainter becomes our representation of them. A similar situation obtains in the case of events temporally distant from our present representational state. In such a case the role, analogous to that of intervening monads, is played by the representational states intervening between the present representational state and the temporally distant one that we unconsciously or confusedly come to represent.

The cardinality of the set of intervening monads is of course always the same, no matter how great or small the spatial distance is between two monads that are represented as spatially positioned. On the other hand, if three monads A, B, C are represented as positioned on the same line, so that B is represented as being spatially between A and C, the set of monads represented as intervening between A and B is a proper subset of the set of monads represented as intervening between A and C.

The monads representationally nearer to a soul are those of its body. If we ignore the semi-physicalistic language Leibniz uses in the following rather long passage from *The Monadology,* where he describes sensing as causal traveling[26] through the phenomenal spatio-temporal plenum, we can see that the interpretation proposed above is correct:

> The nature of the monad being to represent, nothing can keep it from representing only a part of things, though it is true that its representation is merely confused as to the details of the whole universe and can be distinct for a small part of things only, that is, for those which are the nearest or the greatest in relation to its individual monad. Otherwise each monad would be a divinity. It is not in the objects but in the modification of their knowledge of the object that the monads are limited. They all move confusedly toward the infinite, toward the whole, but they are limited and distinguished from each other by the degrees of their distinct perceptions.
> 61. In this respect compound beings are in symbolic agreement with the simple. For every thing is a plenum, so that all matter is bound together, and every motion in this plenum has some effect upon distant bodies in proportion to their distance, in such a way that every body not only is affected by those which touch it and somehow feels whatever happens to them but is also, by means of them, sensitive to others which adjoin those by which it is immediately touched. It follows that this communication extends to any distance whatever. As a result, every body corresponds to everything which happens in the universe, so that he who sees all could read in each everything that happens everywhere, and, indeed, even what has happened and will happen, observing in the present all that is removed from it, whether in space or in time ... But a soul can read within itself only what it represents distinctly; it cannot all at once develop all that is enfolded within it, for this reaches to infinity.
> 62. Thus, although each created monad represents the whole universe, it represents more distinctly the body which is particularly affected by it and of which it is the entelechy. And as this body expresses the whole universe by the connection between all matter in the plenum, the soul also represents the whole universe in representing the body which belongs to it in a particular way. (L, 648-649; G, VI, 617.)

The passage quoted above is interesting for a number of reasons. We can summarize them, thus making clear what, we think, Leibniz's contentions are concerning important features of the nature of monadic representation:

(i) Leibniz states quite clearly, at the beginning of the quoted passage, that the nature of the monad is *to represent.* Every windowless monad is a perspective representor of the world, both synchronically and diachronically.

---

[26] Considered as the byproduct of immediate touching, partly resulting from the non- existence of phenomenal or real vacua.

A monad is characterized by its particular representational structure and it is set in motion, so to speak, by its appetition.

(ii) Each monad represents the whole universe, but it is only a small part of it which the monad represents distinctly. The monad represents the rest confusedly as to its details. This is to be expected, says Leibniz, "otherwise each monad would be a divinity".

(iii) Monads are limited and as individuals differ and are distinguished from one another "by the degrees of their distinct perceptions". We should add here that they are also distinguished from one another by their point of viewish representational structures. The representational structure of each monad is perspective in such a way that to each perspective viewing of the world corresponds at most one monad and, because of the principle of continuity or even the principle of plenitude, to each perspective viewing of the world corresponds *exactly* one monad. The opposite is obvious since each monad has a representational structure which is perspective in a *unique* way. In the passage quoted above, Leibniz does not mention perspective representation as a means for monadic individuation. He does it elsewhere[27]. Yet one could, perhaps, argue that perspective representation is a consequence of the differences of the degrees of the monads' distinct perceptions.

(iv) Representational distinctness or confusedness is connected with the feeling of distance. It is the case, according to Leibniz, that "every motion in this plenum has some effect upon distant bodies in proportion to their distance". Distance is connected with the plurality of monads represented as being in between us and the represented, as distant, object, in a way that the greater the multitude of monads represented as intervening between us and the object (and as we have already said, through the eyes of which monads we come to see it) the fainter its representation by us becomes.

(v) In a sense, the nearest thing to a dominant monad (a soul) is its body. So, "it represents more distinctly the body which is particularly affected by it and of which it is the entelechy". Through its body the soul represents the whole universe. Yet it is not only *nearness* to the soul that characterizes a soul's body. A soul and its body are connected in a particular way which, presumably, has to be cashed out in terms of a special kind of distinctness of

---

[27]Consider, e.g., the following passage from *The Monadology*.

> 57. Just as the same city viewed from different sides appears to be different and to be, as it were multiplied in perspectives, so the infinite multitude of simple substances, which seem to be so many different universes, are nevertheless only the perspectives of a single universe according to the point of view of each monad (L, 648; GVI, 616)

the representations in the soul of the monads which are constitutive of the body; the expression "constitutive of the body" is used in the sense that the monads, which phenomenally constitute the body, are represented in the soul as all together forming it.

In order to develop further the proposal concerning the correspondence of the phenomenal relation of "spatially between" to the representational fact of being "representationally between" and apply it so that the Leibnizian doctrine of infinite divisibility of whatever is spatially extended makes sense, we have to find a way of formalizing it which would make clear its basic ingredients. With that end in mind we can adopt a first order language with equality, "=", the vocabulary of which is given below.

(a) The first group of symbols contains all the logical ones, i.e., the usual connectives; "¬" (negation), "∧" (conjunction), "∨" (disjunction), "→" (implication), "↔" (biconditional), the two quantifier symbols "∃" (existential), "∀" (universal), a left "(" and right ")" parenthesis symbols, and variable symbols "x", "y", "z", ...

The second group contains two subgroups. The cardinality of each one of the subgroups is that of the continuum. A typical symbol belonging to the first subgroup will be of the form "a_r", where r is a real number; a typical symbol belonging to the second subgroup will be of the form "b_p", where p is a real number. The first subgroup of symbols is to provide us with names for monads, the second with names for their representational states.

The third group contains two unary predicate symbols "M" and "R_s"; "M" singles out monads, "R_s" representational states.

At this point some explanatory remarks are necessary. The language we describe looks as though it commits us to a non-Leibnizian ontology. After all, in Leibniz's system we never find monads *simpliciter*. Every monad is always in a certain representational state. Nevertheless, this is not a real difficulty. The term "monad" does not mean here a representor stripped of all of its representational content. Such a representor would no longer be a representor but a mere nothing. On the other hand, an equally unacceptable alternative would be to consider a monad as a certain configuration, or in other words as a mere bundle of representational states. "Monad" means here simply the metaphysical entity that contains its representational states once and for all, so that its priority over them is exactly its own principle of individuation. "Monad," in other words, means the diachronic entity which exists by virtue of the fact that its representational states are being actualized, although it is metaphysically prior to them.

(d) The members of the fourth group of symbols are the infinitely many function symbols "S_1", "S_2", ... "S_n", ... and the binary function symbol "R". For any natural number n the function symbol "S_n" plays the role, when interpreted, of a partial function that, when applied to n monads or representational states, gathers them into an n-membered set. Of course, no sets of monads and representational states form metaphysically real entities in the Leibnizian system. The reasons for having such set forming operators in our language is therefore not to be found in ontological considerations. They are included merely for the sake of convenience.

The binary function symbol 'R', on the other hand, has a special role to play. It signifies a partial function which, when applied to an ordered pair consisting of a monad as its first member and a monad, a representational state, a set of monads, a set of representational states, or a set of monads and representational states as its second member, has as its value the new and unique representational state of the monad which was the first member of the pair representing the second member of the pair.

We can see now how the phenomenal relation of "spatially between" can be expressed in our formal language so that its metaphysical counterpart, (i.e., the representational fact that corresponds to it), also enters the picture. Let us consider again three monads a_1, a_2, a_3. Let us assume that I perceive the monad a_2 to be spatially between a_1 and a_3, and that I (or the dominant intelligent monad that is I) am signified by the symbol "a_4". That a_2 represents a_3 can be expressed by:

$$R(a_2, a_3)=b_{23},$$

where b_{23} is the representational state of the monad a_2 representing the monad a_3. Similarly, we can write:

$$R(a_2, a_1)=b_{21},$$

making the obvious corresponding interpretation. That the monad a_1 represents a_2 as representing a_3 can be expressed by:

$$R(a_1, b_{23})=b_1,$$

where b_1 is the representational state of the monad a_1 representing a_2 as representing a_3. The expression:

$$R(a_3, b_{21})=b_3 ,$$

has again an obvious corresponding interpretation.

The more complex fact that the monad a_2 represents both a_1 as representing a_2 as representing a_3 and a_3 as representing a_2 as representing a_1 can be expressed by:

$$R(a_2, S_2 (b_1, b_3))=b_2 ,$$

where b_2 is the representational state of the monad a_2 that we just described, and $S_2 (b_1 ,b_3)$ is the set of the representational states b_1 and b_3 of the monads a_1 and a_3 that we described above. Finally, the fact that I perceive the monad a_2 to be spatially between a_1 and a_3 is due to my internal viewing of my representational state, which can be expressed as follows:

$$R(a_4 ,S_3 (b_1 ,b_2 ,b_3))=b_4 ,$$

where $S_3 (b_1 ,b_2 ,b_3)$ is the set of the representational states b_1, b_2, b_3 (of the monads a_1, a_2, a_3 respectively) which taken together form what we called the representational fact of a_2 being representationally between a_1 and a_3. This representational fact is something that obtains in the real (metaphysical) realm. My perceiving of monad a_2 as spatially in between a_1 and a_3 is my representing of the representational fact of a_2 being representationally between a_1 and a_3. That I represent such a fact is not my exclusive privilege. If b_1, b_2, and b_3 obtain, then every monad a_r which is not identical[28] to either a_1 or a_2 or a_3 perceives a_2 as spatially in between a_1 and a_3 by being in a representational state similar to mine and described by the expression:

$$R(a_r, S_3 (b_1, b_2, b_3)) = b_r .$$

What we called a representational state in the course of the above analysis is only a part of a complete representational state of a monad. This complete state is a static (momentary) perspective representation of the whole world of monads as also in complete and simultaneous representational states. Additionally, individual characteristics of these states depend on their different perspective structures, as well as on their confusedness or distinctness as

---

[28] a_1 or a_3 represents a_2 as spatially between itself and a_3 or a_1 by being at the representational state b_1 or b_3 respectively. a_2 represents itself as spatially between a_1 and a_3 by being at the representational state b_2.

wholes, or the relative confusedness or distinctness of their parts. Yet these facts do not have any particular bearing upon the analysis of the phenomenal relation of "spatially between" we have offered. Whether or not a monad distinctly perceives such a relation as phenomenally holding between any three monads a_1, a_2, a_3 ,does not change the fact that this same monad, consciously or unconsciously, distinctly or confusedly[29], will represent faithfully whatever obtains in the real world of monads representing one another. To carry out an analysis, such as the one given above, one has to resort to a level of abstraction at which irrelevant details do not matter.

That whatever is spatially extended is not only infinitely divisible, but in fact infinitely divided is one of the central doctrines of Leibniz's system. The phenomenal world, in all its aspects, is continuous. It is grounded at the same time, by being its faithful representation upon the real world of monads and upon their representational states. The real world of monads is in one sense discrete. In order to see how a continuous phenomenal world is based upon what appears to be the discreteness of what is metaphysically real, we have to account at least for a basic characteristic of spatial continuity, namely, density or infinite divisibility of the spatially extended, in terms of the previously discussed phenomenal relation of "spatially between" and its metaphysical correlate.

What are the basic ingredients of such an account? One can start by explaining what it is to be infinitely divisible. An extended phenomenal object such as a slim wooden rod can be cut, at least in principle, *anywhere* along its length. First we must determine how to understand "anywhere." For a continualist like Leibniz, "anywhere" would mean that whatever cut we perform, the resulting pieces are always able to be cut also. A modern, mathematically sophisticated thinker would interpret such a claim as purporting to affirm that "anywhere" means at *every point.* Such an interpretation is problematic for the simple reason that points are problematic entities. They are useful fictions for getting a Cantorian mathematical theory of the spatio-temporal order, but they are not to be found anywhere in the world of extended objects. Whatever exists in a world of extended objects is extended and therefore extensionless points are mere *entia rationis*. In his mature years, Leibniz was aware of this problem and his analysis of the idea of infinite divisibility was based upon his *non-spatial* substantial units, i.e., his monads representing one another as representing. Each monad carries with it a representational structure which together with the representational structures

[29] And always perspectively.

of all the monads in the world forms the continuous metaphysical substratum of representational facts. This metaphysical susbtratum, as faithfully represented, constitutes the realm of the well-founded continuous phenomena.

A modern mathematician of the Cantorian breed is so used to constructing the real line out of points that when he comes to the application of his continuous mathematics to the spatio-temporal realm, *he very seldom realizes* the dubious ontological character of points. When asked what a point is he will either answer with an idealization, such as "a point is the limit of a certain process", or by using the idea of something comparatively so small or so distant from us that its dimensions can be ignored. This lack of realization extends to ignorance of the metaphysical puzzle, which cannot be solved in the framework of a mathematical theory, of how one can construct an ontology of spatially extended objects out of unextended substantial units. Leibniz, profoundly disturbed by this puzzle, was forced to dismantle the physical continuum in such a way that the points became, in a sense, the metaphysical, non-spatial carriers of the relational structure of the extended objects and the extended objects phenomena *bene fundata*. A basic characteristic of such a structure, according to Leibniz, is that the phenomenal world of extended objects is a continuous plenum and, therefore, everywhere dense. But how can one make sense of such a contention using only tools from Leibniz's stock? In order to give a metaphysically sound account of what it means that the phenomenal world of extended objects is everywhere dense, we propose to use the basic idea developed in this section. This is the idea according to which a representor a_r perceives a monad a_2 as being spatially between the monads a_1 and a_3 whenever it represents monad a_2 as being representationally between a_1 and a_3. Such a perception is a well-founded phenomenon if a_2 is in fact representationally between a_1 and a_3 ; that is, if in the real world of monads the representational states b_1, b_2, and b_3 of the monads a_1, a_2, and a_3 respectively, obtain simultaneously with the representation of b_1, b_2, and b_3 by the representor a_r. We can now classify the basic ingredients of our answer to the question of how the phenomenal corresponds to the real in the case of the density of the world of the spatially extended. Our intention is to show that the Leibnizian contention, according to which the phenomenal spatial world is everywhere dense, makes good metaphysical sense.

(a) The contention that the phenomenal spatial world is everywhere dense can be based upon representational facts which give rise to a universal statement concerning what really obtains at the metaphysical level of monads as they represent one another, namely, the statement "for every two distinct monads x_1 and x_2 there exists a third monad distinct from them, x_4, which is representationally between them."

(b) Every representor, whether aware of it or not, represents all those representational facts so that the universal statement "the phenomenal spatial world is everywhere dense" can be understood as "for every two distinct monads x_1 and x_2, and every other monad, x_3, distinct from them, there exists a monad x_4, distinct from the previous three, such that x_4 is representationally between x_1 and x_2, and x_3 represents x_4 as representationally between x_1 and x_2.

Using (a) and (b) we can at last formalize the Leibnizian contention that the phenomenal spatial world is everywhere dense (as based upon representational or, so to speak, metaphysical density) as follows:

$$\forall x_1 \forall x_2 \forall x_3 (M(x_1) \wedge M(x_2) \wedge M(x_3) \wedge x_1 \neq x_2 \wedge x_1 \neq x_3 \wedge x_2 \neq x_3 \rightarrow \exists x_4 \exists x_5 \ldots \exists x_{10} (M(x_4) \wedge R_S(x_5) \wedge R_S(x_6) \wedge \ldots \wedge R_S(x_{10}) \wedge R(x_4, x_1) = x_5 \wedge R(x_4, x_2) = x_6 \wedge R(x_1, x_6) = x_7 \wedge R(x_2, x_5) = x_8 \wedge R(x_4, S_2(x_7, x_8)) = x_9 \wedge R(x_3, S_3(x_7, x_8, x_9)) = x_{10}))$$

Before moving on to an examination of time, phenomenal change, and monadic change, a few remarks about this model and formula are necessary.

(1) The quantifiers used range over either monads or monadic states. That is, they range over *really* existing entities or states of them. A reminder is necessary here. The term "monadic states" means the appropriate representational parts (of complete monadic states) which are relevant to the construction of the proposed model. A complete representational state of a monad is a static, momentary, and perspective representation of the world of monads as they represent one another, and as such it is characterized by a complexity only part of which is relevant to the representation of the real world of monads as representationally everywhere dense.

(2) The variables x_1, x_2, x_3, x_4 refer to monads and the variables x_5, ... x_{10} refer to monadic representational states. The representational state x_7 is that of the monad x_1 representing x_4 as representing x_2. x_8 is the representational state of the monad x_2 representing x_4 as representing x_1. Those three states taken together form the representational fact described by the phrase "x_4 is representationally betweeen x_1 and x_2". This is a fact obtaining at the real, metaphysical level. Every monad represents it. The monad x_3 perceives x_4 as being spatially between x_1 and x_2 by being at the representational state x_{10}, that is, by representing the representational fact of the monad x_4 being representationally between x_1 and x_2.

(3) In formulating the contention that the phenomenal spatial world is everywhere dense, as based on the contention that the real world of monads is

representationally everywhere dense, we used a monad x_3 distinct from x_1, x_2, and x_4. This does not present a problem for the model. If x_3 were identical either with x_1 or x_2 or x_4, we would say that x_1 or x_2 or x_4 represents x_1 spatially between x_1 and x_2 by being at the appropriate representational state x_7 or x_8 or x_9[30].

C. TIME, PHENOMENAL CHANGE, AND MONADIC CHANGE

Leibniz usually defines time in juxtaposition to space. He maintains that both are innate abstract representations or orders pertaining to coexistence and succession, or to existence which is not simultaneous. Consider the following short passage from a letter of Leibniz to Des Bosses, dated June 16, 1712:

> ... space is the order of coexisting phenomena, as time is the order of successive phenomena. (L, 604; G, II, 450)

Two years later, in a letter of his to Nicolas Remond, dated March 14, 1714, Leibniz states that:

> ... space is exactly the same as the order of coexistence, as time is the order of existence which is not simultaneous. (L, 656; G, III, 612.)

One can find a similar passage in Leibniz's "Third Paper" (*The Leibniz-Clarke Correspondence*). In this passage he stresses the relative character of space and time and he continues making clear that *coexistence* and *succession* are their corresponding basic features:

> 4. As for my opinion, I have said more than once that I hold space to be something merely relative, as time is: that I hold it to be an order of coexistence as time is an order of successions. (A, 25; G, VII, 363.)

Leibniz holds time to be an ideal entity[31]. That is, it is an innate idea which is not produced as an extension out of our experiential awareness of finite durations. Additionally, it does not correspond to anything unitary and substantial. Time, in other words, is not an idea which corresponds to an absolute container of change existing by itself, and as such it lacks a referent which would qualify as a *thing* in its own right. It can be thought of as the order of relations of succession in phenomenal change when abstracted from the particularities of their own relata. It is in this sense that Leibniz considers time to be relative or relational. Time is an innate idea which we come to realize that we possess by paying heed to it through the observation of

---

[30] This case has already been mentioned in footnote 28 in this section.

[31] e.g., A 64; G, VII, 396 or GM, VI, 247, where he maintains that " ... space, time, and motion each are in some degree an *ens rationis*."

phenomenal change. As such it conforms to the phenomena by being the abstract representation of the relational structure of the phenomenal change itself.

According to Leibniz, time, since it is an ideal entity, belongs to the realm of mathematical concepts[32]. More specifically, it can be characterized as one-dimensional, directional and ideal continuum. As such, it possesses the following properties:

(a) Linearity,
(b) Directionality,
(c) Uninterruptedness, and
(d) Potential density[33] or potential infinite divisibility.

Time, that is, can be pictured as a straight line, prior to its parts or points, upon which a definite direction has been imposed. Such a line is potentially infinitely divisible anywhere throughout its length, so that between any two actualized divisions a third one can be actualized too. Its priority over the points or its parts entails that it is not constituted by them. There are no actual points or parts contained in it. Such a line is continuous not only in the sense that it is potentially dense, but also in the sense that it is a unitary, uninterrupted ideal whole.

As a mathematical concept, time conforms to what is actual by being the abstract representation, or "the order" of successive phenomena. Truths about its structure not only cannot be violated by well-founded phenomena-and more specifically by phenomenal change-but, on the contrary, govern them by being eternal truths about such change (when abstracted from its particularities.) On the other hand, phenomenal change cannot be equated with time. Time (directionality excluded) is uniform or homogeneous, phenomenal change is not[34]. Needless to say, directionality could be thought of as violating uniformity or homogeneity. Instants, no matter whether they are actual or potential, differ from points because of considerations pertaining to directionality. That is, an instant A can be in a unique and exclusive way either

---

[32] See NE, 110; G, V, 100.

[33] A continuum, according to the modern point of view, is characterized by sequential completeness as well, something that Leibniz could not have been aware of.

[34] The following passage from Leibniz's "Fifth Paper" (*The Leibniz-Clarke Correspondence*) is quite illuminating:

> 27.The parts of time or place, considered in themselves, are ideal things; and therefore they perfectly resemble one another like two abstract units. But it is not so with two concrete ones, or with two real times, or two spaces filled up, that is, truly actual. (A, 63; G, VII, 395.)

before or after an instant B. Consider the following quotation from a letter of Leibniz to Louis Bourguet, dated August 5, 1715:

> I admit however that there is this difference between instants and points - one point of the universe has no advantage of priority over another, while a preceding instant always has the advantage of priority, not merely in time but in nature over the following instants.. (L, 664; G,III, 581,582)

The ideality of time is connected, for Leibniz, with its uninterruptedness, that is, with the non-existence of point-instants in the one-dimensional, directional, ideal continuum, which is time. Time as an ideal entity does not contain instants. Time does not consist of instants, whereas phenomenal change does. Time, is uniform and uninterrupted. In a passage from the "Fifth Paper" (*The Leibniz-Clarke Correspondence*) Leibniz says:

> Nothing of time does ever exist, but instants; and an instant is not even itself a part of time. Whoever considers these observations will easily apprehend that time can only be an ideal thing. (A, 72-73; G, VII, 402.)

We should note here that Leibniz's terminology is slightly misleading. He runs together his notion of time as an ideal entity with the notion of phenomenal change.

Instants, however, are not parts of time. But they are not its ultimate constituents either, because time, an ideal entity, does not consist of instants which are to be understood as actual momentary states of the world of successive well-founded phenomena. They are, on the contrary, the constitutive ultimate elements of phenomenal change or of actual duration, in a way which is analogous to the composition of an extended object out of the representations of the monads which, we can catachristically say, compose it. In short, instants, with all their particularities, as momentary states of the world of well-founded phenomena constitute phenomenal change which is actual and not time, which is ideal.

Phenomenal change or parts of it, namely phenomenal durations, are the structural equivalents of the physical continuous universe or, correspondingly, of parts of it, i.e. of spatially extended chunks, such as extended objects. If we were to consider their most basic similarities and dissimilarities in some detail, we would come up with the following list.

1. A specific phenomenal duration is *actually* dense or *actually* infinitely divisible, resembling in that respect an extended object. That is, phenomenal change consists of instants which are densely interwoven in such a way so that between any two instants A and B, there is always an instant C which comes after A and before B. Actual infinite divisibility of phenomenal change (as well as actual infinite divisibility of the spatially extended), is a structural

characteristic of an architectonic nature which can be thought of as stemming from the principle of continuity.

2. Both phenomenal change and the spatially extended do not qualify as genuine continua, because they are not substantial unitary wholes, but continuously structured extended aggregates of indivisible unextended ultimate units. Phenomenal change consists of instants which, we shall see, are representations of momentary states of the world of monads representing one another. The spatially extended, at a given instant, is the representation of an infinitude of monads. As such it consists of an infinitude of unextended ultimate units (each one of which is the representation of a specific monad) interwoven into one representation so that they appear as a three-dimensional continuous whole. Both phenomenal change and the phenomenally spatially extended do not qualify as genuine continua, according to Leibniz, because they do not satisfy the property of uninterruptedness. Both have actual parts, and, though continuous, they both can be thought of as discretely structured, i.e., as consisting of ultimate indivisible units. Genuine continua, according to Leibniz, do not have parts. There is, of course, an obvious dissimilarity. No phenomenal duration exists all at once, although all of the constitutive elements of the spatially extended at a specific moment do. All that ever exists of phenomenal change are single instants following one another in the unfolding of the history of the world of phenomena. By being in an instantaneous representational state we represent perspectively all the past and future history of such a world; but then, such a representation is a static viewing of it from a particular momentary "now." This "now" is, so to speak, the perspective temporal viewing center of the representation. More specifically, if we assume that we are at a particular momentary representational state A, we represent ourselves as having been or as going to be at particular momentary representational states in the past or future. Such indirect temporal representations of past or future representational states as they exist in one and the same representation of a particular "now"-whether or not we are conscious of them-form a continuum, i.e., they at least satisfy density.

3. Both the physical universe, considered as the totality of the spatially extended at any particular moment, and phenomenal change, considered as the totality of all of these phenomenal moments, are unbounded. There is no first fundamental instant in phenomenal change and no last instant either[35].

---

[35] Nor, for that matter, any first or last instant in monadic change.

Consider, for instance, the following interesting passage from a letter of Leibniz to Louis Bourguet, dated August 5, 1715:

> As for the nature of succession, where you seem to hold that we must think of a first, fundamental instant, just as unity is the foundation of numbers and the point is the foundation of extension, I could reply to this that the instant is indeed the foundation of time but since there is no one point whatsoever in nature which is fundamental with respect to all other points and which is therefore the seat of God, so to speak, I likewise see no necessity whatever of conceiving a primary instant. (L, 664; G, III, 581.)

In the passage quoted above Leibniz argues that there is no "necessity ... of conceiving a primary instant" on the ground that there is an analogy between points and instants. Since there is no one single privileged point, there is no one privileged instant to serve as a primary one. In the following passage from Lebniz's "Fifth Paper" (The Leibniz-Clarke Correspondence), he suggests indirectly that there is no last instant either:

> The case is the same with respect to the destruction of the universe. As one might conceive something added to the beginning, so one might also conceive something taken off towards the end. But such a retrenching from it, would be also unreasonable. (A, 76; G, VII, 405.)

4. The first obvious dissimilarity between the spatially extended and phenomenal change concerns the notion of directionality. In what appears as spatially extended there is not, at least in principle, a privileged direction or even a privileged set of directions as a proper subset of the totality of possible directions. A monad M can be represented as moving from a position A to a position B (positions considered with respect to a referential system consisting of the representations of other monads as in fixed positions) or *vice-versa*, provided that these representations can qualify as well-founded phenomena. Phenomenal change, on the contrary, is irreversible and therefore directional. We cannot move at will from the future to the past. Appetition is the continuous striving toward the future. We contain in our individual concept our personal history once and for all (which is a perspective viewing of the world's history), but we cannot decide to live it by choosing at will or at random certain portions of it (the happier ones, for instance), or by building up a personal time machine which would enable us to go back to the past or to travel forward so that we can see the future with the eyes of the present. In other words, we cannot have either flash-backs or shortcuts to what is temporally in store for us.

5. A second basic dissimilarity between the spatially extended and phenomenal change arises from dimensionality considerations. The totality of

the spatially extended, at any particular moment, forms a three-dimensional physical and, therefore, actual continuum. Phenomenal change, on the other hand, is a linear or one-dimensional actual continuum.

A three-dimensional physical continuum is abstractly represented by the ideal three-dimensional continuum of Euclidean geometry. Phenomenal change, however, is abstractly represented by a directional straight line. One of the most interesting by-products of the dimensionality discrepancy concerns our epistemological access to the existence of vacua in them. That there are no vacua in any order of possibles or actuals in this world is a central Leibnizian doctrine. Yet, such vacua as indicating changeless durations, it would not be possible to be empirically verified as existing even if phenomenal change contained some. How could they be verified, if the only way to do so were either by experiencing change in them, or by measuring change or its absence within something of a temporal nature which would run parallel, but independently of it? In the *New Essays on Human Understanding* we find the following interesting passage:

> I will add a comparison of my own to those that you have given between time and space. If there were a vacuum in space (for instance, if a sphere were empty inside), one could establish its size. But if there were a vacuum in time, i.e., a duration without change, it would be impossible to establish its length. It follows from this that we can refute someone who says that if there is a vacuum between two bodies then they touch, since two opposite poles within an empty sphere cannot touch-geometry forbids it. But we could not refute anyone who said that two successive worlds are continuous in time so that one necessarily begins as soon as the other ceases, with no possible interval between them. We could not refute him, I say, because that interval is indeterminable. If space were only a line and if bodies were *immobile*, it would also be impossible to establish the length of the vacuum between two bodies. (NE, 155; G, V, 142.)

On the other hand, if vacua in space existed, they would be verifiable by experience, because a vacuum in one dimension could be empirically accessible, at least in principle, by an inspection via the other dimensions, i.e., by moving, as it were, all around such a vacuum. What is interesting is that Leibniz holds that the non-existence of an epistemological access to temporal vacua is characteristic of phenomenal change's *linearity* only, and has nothing to do with its other properties. In the case of a one-dimensional spatial world, if vacua existed, they would not be empirically traceable or recognizable either. That is, linearity of such a world would necessarily imply the epistemological inaccessibility of spatial vacua.

6. A third dissimilarity is related to the following fact: the totality of the spatially extended at a specific phenomenal moment is a static representation of the world of monads representing one another at the metaphysical moment which corresponds to this specific phenomenal moment. On the other hand, phenomenal change, as a series of representations of real or monadic change, is of a dynamic or at least kinematic nature. Such change is represented at any particular moment in the framework of a static representation, wherein we can distinguish the present from the past and represent unconsciously the future.

What we see or experience as change is the representation of real or monadic change. Before we go on to examine three alternative models for monadic change, the final one of which we take to be the correct one for interpreting Leibniz, we should address ourselves to the following problems. First, we should specify the nature of simultaneity and accordingly discuss briefly the measurement of change at the phenomenal level. Second, we should distinguish intra-monadic from inter-monadic change. It is our view that intra-monadic change does not correspond to what we would currently call private time. Rather, it incorporates both private and public, or intersubjective time. Inter-monadic change, on the other hand, is the real or metaphysical counterpart of phenomenal change. Our epistemological access to it is not characterized by its immediate grasp by the mind. We come to postulate it (or, rather, Leibniz does) as the firm objective basis for explaining the coherence of the enfolding series of phenomenal change.

Simultaneity, at the phenomenal level, can be better understood if we use the simple example of two monads A and B representing each other as follows: a monad can, in principle, distinguish in a momentary representation-whether or not such a representation as a whole is confused or distinct, whether or not the monad is conscious of it-features which can be described by using indexicals such as "now," "before," and "after." Monad A, by being in a representational state R_A, represents B as being now in a particular state R_B. In other words, A represents B as being in the representational state R_B neither before nor after is itself in the representational state R_A. A represents B as being at the representational state R_B precisely when it itself is in the representational state R_A. Representing something as taking place *now* (as opposed to before or after) can be cashed out in terms of its special relative distinctness or lack of confusion, or even through its directness as a representation. For example, memory[36] is a species of indirect representation

[36] Consider the following passage from *The Monadology*.

according to which, we represent ourselves as having represented something which was directly represented as happening at what was then not past but present[37].

There are now three cases we must consider.

(a) Let us assume that A, by being in representational state R_A, represents B as being now in a particular representational state R_B. Let us also assume that B, by being in R_B, does not represent A as now being in the representational state R_A. Such a case is excluded because it obviously violates Leibniz's doctrine of pre-established harmony.

(b) By being in the representational state R_A, A represents B as being now in representational state R_B, and B, by being in R_B represents A as being in the representational state R_A. From God's point of view, both A and B agree as to the events that are simultaneous for them, simply because their representational states are appropriately coordinated in the way already described. But even this is not enough to give us the desired objectivity with regard to simultaneity. For this, phenomenal simultaneity has to be based upon monadic reality itself and, more specifically, upon simultaneity at the level of the real. The mere pre-established harmonious coordination of appropriately correlated monadic or representational states is a solid basis for intersubjectivity, but not for metaphysical objectivity.

(c) To complete the picture one must say that phenomenal simultaneity is based upon real or metaphysical simultaneity. Leibniz considers only monadic change as real change, i.e., the movement from a representational state to future ones. Monadic change is what is correctly represented as phenomenal change. But monadic change is something that characterizes and affects every monad. Should we adopt the view that monadic changes, as pertaining to different monads, are not correlated appropriately at the metaphysical level? Or should we adopt the more radical view that simultaneity at the level of the real is meaningless? The model we described in (b) works pretty well, after all, for simultaneity at the phenomenal level, without any extra metaphysical

---

26. Memory provides a kind of consecutiveness to souls which simulates reason but which must be distinguished from it. Thus we see that when animals have a perception of something which strikes them and of which they have had a similar perception previously, they are led by the representation of their memory to expect whatever was connected with it in this earlier perception and so come to have feelings like those which they had before. (G, VI, 611.)

[37] We should not confuse *memory* with *apperception*. We remember something by representing ourselves as having represented this something directly. On the other hand, we apperceive something if we represent ourselves as directly representing this something.

assumptions. Our experience of the world of well-founded phenomena can be temporally quite coherent without having to commit ourselves to simultaneity at the metaphysical level. But there is, we think, good reason for feeling uneasy about such a solution. Leibniz's theory of truth is not a coherence theory but a representational one. Phenomena are well-founded neither because they form a coherent pattern, nor because intersubjectivity can be established independently of the desired synchronicity of the appropriate representational states at the metaphysical level. They are well-founded because they are correct representations of the metaphysical reality. Consequently, what appears as simultaneous has to correspond to what really is simultaneous. In the case of our previous example of monads A and B, for instance, A and B should be in representational states R_A and R_B which are metaphysically simultaneous. Phenomenal simultaneity then would be a well-founded phenomenon based upon the reality of metaphysical simultaneity.

This discussion is immediately related to our worries about intra-monadic and inter-monadic change. The model we presented in (a) calls for a world in which intersubjectivity, with respect to simultaneity, is violated. Phenomenal change is experienced privately[38] in such a way that from God's point of view there is no harmonious agreement between the representations of the different representors. A monad could, of course, get the impression that intersubjectivity obtains, especially if its representations were appropriately coherent. On the other hand, the model we presented in (b) calls for a world in which intersubjectivity is guaranteed. It is nonetheless true that such intersubjectivity is not a good substitute for what really obtains in the realm of monadic change. That is, intersubjectivity is well-founded if there is only one *inter-monadic time-like order*, i.e., if that which intersubjectively appears as simultaneous is the correct representation of what is really or metaphysically simultaneous.

According to the position we developed, there is an important difference between the metaphysical correlate of the spatially extended and monadic change. We think that although Leibniz could successfully maintain the non-reality of space (in the sense that monads are not positioned in a metaphysical spatial absolute container) he was mistakenly holding a similar position regarding time. Time and space are ideal entities for Leibniz and, in that sense, they could be thought of as being non-real in a similar way. The spatially extended and phenomenal change are also on a par in the sense that they are well-founded appearances. But their well-foundedness is based upon a

---

[38] We do not use the term "private" as involving connotations of a pyschologistic nature, but mean by it simply the personal, unshared access one has to one's own representations.

monadic reality which, although not contained in a metaphysical, spatial and absolute container, changes in an absolute sense by moving from state to state so that metaphysical simultaneity is as real as it can be. Such real change is a prerequisite for the well-foundedness of phenomenal change. That is, what is *phenomenally* now or will be later or was before, has to correspond to what is *metaphysically* now or will be later or was before. In short, monadic change is not merely ideal thing or a phenomenal time-like order, but a real time-like order. Monadic change relations are real, connecting together metaphysical past and future, in a way that metaphysical spatial relations are not. Phenomenal spatial relations are well-founded not because they are representations of metaphysical spatial relations, but because they are specific representations of metaphysically simultaneous representational states of monads representing one another in a pre-established, harmonious way. Metaphysical spatial relations are not to be found in the world of monads. However, because monadic change is real, monadic change relations are to be found in the world of monads. Monadic change is real, not in the sense that it exists all at once, but in the sense that metaphysical temporal specifications are already a part of the world of monads as they really change and not as they merely represent one another as changing.

Measurement of phenomenal change is based upon Leibniz's idea of simultaneity or simultaneous existence. Measurement presupposes synchronicity of different processes and synchronicity presupposes simultaneity of the corresponding representational states of the monads involved in these processes. Measurement of phenomenal change is always the result of the comparison between a certain phenomenal duration — which we intended to measure — and a temporal unit of measurement, that is, the duration of a certain process which we have good reason (scientific or empirical) to believe to be of a constant temporal length. Ordinary or technologically sophisticated clocks have the function of helping us to measure phenomenal change in accordance with such a rationale.

Leibniz thought of the number of intervening states as indicating the time that elapses between two phenomenal states of the world, for example states A and B[39]. But this is not entirely correct, because we know by now that two line segments of unequal lengths are order isomorphic, which means among other things, that they have the same cardinality. The way out of this problem would be to consider the set of states between A and B, and not the number of the intervening ones, as indicating the time length between them. Then, if we had

[39] See, e.g., A, 89-90; G, VII, 415 or GM, VII, 18.

three momentary states of the world A, B, and C, such that A was before B and B was before C, the set of states intervening between A and B and the set of states intervening between B and C would be non-intersecting proper subsets of the set of states intervening between A and C. Moreover the union of the sets of states intervening between A and B and between B and C - the state B included - would be the set of states intervening between A and C. But even this would not get us entirely out of the difficulty of comparing durations which have an empty intersection. Additionally, set-theoretic inclusion comparisons are not a good substitute for metric comparisons. For such comparisons to be possible, one should presuppose metric and not simple set-theoretic structures. A possible solution to this problem could be found if we were ready to abandon thinking of lengths of phenomenal durations (or of lengths in general) as being exclusively dependent upon the number or the set of intervening units. By doing so we would start thinking instead about the temporal distance of a phenomenal state A from a phenomenal state B as already determined by characteristics of the representations themselves. For instance, if we were in the metaphysical state that corresponds to the phenomenal state of the world A and we were representing the real state corresponding to B as B, then we would represent it so that such a representation involved already a predicate specifying the temporal distance between A and B.

As we have said, what we have called monadic change corresponds to phenomenal change. Such real change is the metaphysical foundation of phenomenal change, that is, phenomenal change is the correct representation of monadic change. In what follows we will examine their interconnection, especially as referring to density. We will propose a model which, we think, will make clear the nature of the essential isomorphism that obtains between them. But we will first examine two other models of monadic change for the purpose of creating the appropriate contrast between them and the one we think to be the correct interpretation.

1. *The Whiteheadian Model*[40]

According to this model, monadic change could consist of ultimate units of a durational nature. That is, a monadic state could be not a point-like entity, but an entity resembling a line segment. In such a case two possibilities are open to us:

---

[40] We give the model this name because of its resemblance to the Whitehead's model of forming the world out of durational and not instantaneous perceptual units. See, e.g., [148], esp. Chapter IV, or [150], esp. Part III.

(a) We could think of a monadic state as durational at the metaphysical level and as non-durational at the phenomenal one. In other words, we could imagine the internal viewing by a monad of such a state as characterized by phenomenal synchronicity or simultaneity; that is, we could imagine monadic change as consisting of durational monadic states which are represented, and therefore appear to us, as momentary. Such a case is ruled out as an interpretation of Leibniz because it implies that our representation of monadic change (that is, phenomenal change) is not a correct mirroring of it. We would represent it in a fundamentally incorrect way, since we would view metaphysically lengthy entities as point-like ones. The parallel of such a situation in the case of the spatially extended would be to accept the view that monads, though metaphysically extended-whatever that means-are represented as spatially unextended. If we were not to exclude such a case as a model of the Leibnizian metaphysics of change, we would admit the absurd thesis that this world was created by a God who likes to play a game of perceptual deception with his creatures. We would have to admit that God is a deceiver, since he has created a world of representors who represent in a fundamentally incorrect way what really or metaphysically obtains.

(b) The second possibility would be to adopt the view that, according to Leibniz, monadic states are not only durational, but that they are also represented as such. That is, phenomenal change, as a correct[41] representation of monadic change, consists of ultimate units which are also of a durational nature. We could additionally adopt one of the following three alternatives as expressing what Leibniz believed with regard to metric comparisons between monadic states: (i) The duration of any such state is equal to the duration of any other one. (ii) The durations of two monadic states are not necessarily equal, but there is nonetheless a lower bound of their lengths. The lengths of these durations, in other words, can differ, but they cannot be smaller than a certain fixed length. (iii) The durations of two monadic states are not necessarily equal and there is no lower bound concerning their length. In this case we could define a time instant as a theoretical or ideal entity using a procedure analogous to the Whiteheadian method of *extensive abstraction*. An *abstractive set* of lengths of durations is any set of lengths of durations which possesses the two properties: 1) of any two members of the set, one contains the other as a part, and 2) there is no length of duration which is a common part of every member of the set. An abstractive set A *covers* another abstractive set B when every member of A contains as its parts some members

[41] A correct representation could be thought of as, e.g., an isomorphic one.

of B. If an abstractive set A covers another abstractive set B, and B covers A, we say that A and B are *mutually* covered. The relation of being mutually covered is an equivalence relation. A *time instant* is then the equivalence class of all abstractive sets which are related to any given one of themselves that way.

No matter which one of these three alternatives we might consider as appropriate, we would be forced also to admit that phenomenal durations, as corresponding to the monadic ones, would satisfy similar conditions with respect to their metric comparisons. Independently of this point, however, even this second durational model, despite its obvious advantages over the first one, cannot be considered as corresponding to a correct interpretation of the Leibnizian notion of change. Though Leibniz in most places runs together phenomenal and real or monadic change, he is at least clear and definite concerning their composition. *Their ultimate units are momentary and therefore non-durational.* "Nothing of time does ever exist" says he "but instants" (A, 72-72; G,VII, 402).

2. *The Discrete and Discontinuous Model*

In quite a few passages Leibniz talks about monadic change using a mode of speech which could lead one to form the impression that such change is considered by him to be of a discrete and discontinuous nature. Consider, e.g., the following passages[42] from his Correspondence with Arnauld:

> The proposition which has occasioned this discussion is, I may add, very important and merits a firm proof, for it follows that every individual substance expresses the entire universe after its own manner and according to a certain relationship, or, so to speak, according to the point of view from which it looks at the universe; and that its succeeding state is a sequel (although free or contingent) of its preceding state, as though only God and it existed in the world. (M, 64; G, II, 57.)

> By the concept of substance or complete being in general, which implies that its present state is always a natural consequence of its preceding state, it

[42] Or the following ones, which are from the *Clarification of the Difficulties which Mr. Bayle has found in the New System of the Union of Soul and Body* and *The Monadology*

> ... the present state of each substance is a natural result of its preceding state (L, 495; G, IV, 521.)

> ... every present state of a simple substance is a natural consequence of its preceding state, in such a way that the present is great with the future. (L, 645; G, VI, 610.)

> follows that the nature of every individual substance and consequently of every soul is to express the universe. (M, 146; G, II, 113.)

> ... every present state of a substance is a consequence of its preceding state. (M, 148; G, II, 115.)

If we were to adopt the view that, according to Leibniz, monadic change is discrete and discontinuous, we would be faced with the problem of reconciling it with the definite Leibnizian view that phenomenal change is discrete and continuous. It is not clear how this reconciliation could be achieved. The basic idea would be to think of both monadic and phenomenal change as consisting of ultimate non-durational units (of different sorts, of course) which constitute them in different ways.

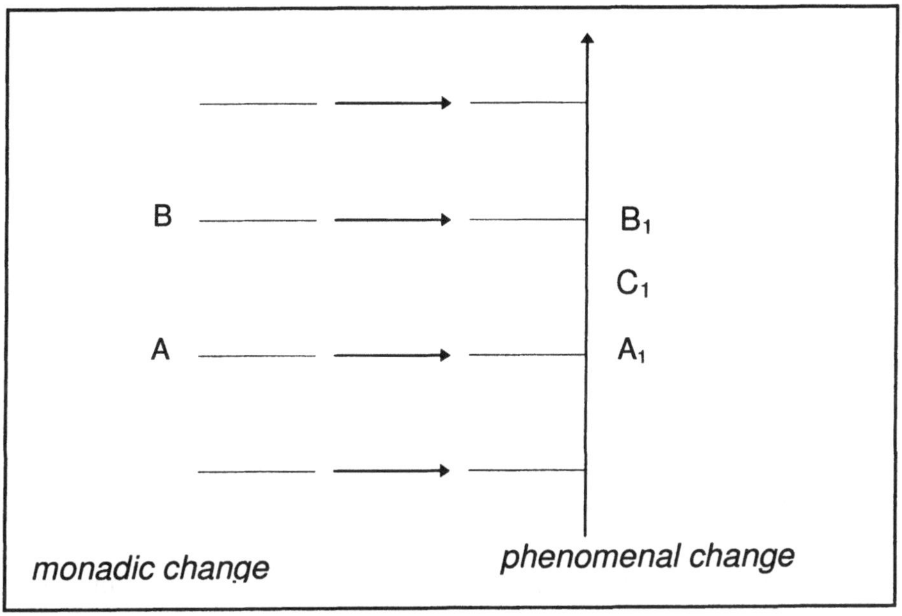

A monad M moves from a preceding state of its own, A, to a subsequent state, B, without passing through any intermediate states. By being in such a metaphysically momentary state, M represents change as if it were continuous. That is, by being in representational state B, M represents change as if densely arranged intermediate states of its own had existed between A and B[43]. To make the picture clearer we could imagine phenomenal change as a directional

[43] We could even talk about states of the world of monads A and B, in which case all monads would move in a metaphysically synchronous way from A to B, viewing the same world perspectively.

straight line and monadic change as a directional discrete and discontinuous series of points running parallel to the line.

To the monadic or representational state B corresponds a representational viewing of the world B_1 and to A, A_1. When monad M is in the representational state B, it represents the world as if it were presently at the phenomenal state B_1. Such a representation contains, of course, representings of what is, or appears to be, the future or the past as well. The representational state A, whose immediate or direct representation is the phenomenal state A_1, is represented as temporally prior to B in B_1. But additionally in B_1 there appear to be states represented as temporally in between A and B, although there were no such real states of monad M. It is as though a continuous tape is running inside a monad, with the monad itself really moving from a momentary state to a momentary state discretely and discontinuously, so that although it does not go through all of the intervening states corresponding to the momentary frames of the tape, it remembers them or, better, represents them in their continuous temporal entirety.

We think a model like this one cannot provide an accurate picture of Leibniz's metaphysics of change, for the following reasons.

(i) The seeming textual support for it, although abundant, is misleading. The mode of speech Leibniz uses in talking about a subsequent state of a monad and its[44] preceding state is sloppy, and it originates in a common way of talking about causal connections between what takes place before and what comes immediately after as a causal consequence of it. Causal connections are commonly phrased in terms of ordered pairs of events or of states of the world, with a tendency to ignore or put aside the difficulties presented by causal connections via a continuous (i.e., at least dense) string of intermediate states.

In the case of the Leibnizian system, there is enough room to accommodate both phenomenal causality, as determined by the lawfulness of the phenomenal or physical world, and be-causality (or metaphysical causality), as, on the one hand, playing the role of the metaphysical foundation of phenomenal causality and, on the other, being the directional connecting linkage of what metaphysically comes before and what follows after. But this distinction between phenomenal and metaphysical causality cannot be used as the supporting ground for the conjunctive idea of a discrete and discontinuous monadic change and of a discrete and continuous phenomenal one. On the contrary, continuous causal connections at the level of the phenomena have to

---

44 "... son estat precedent..." (G, IV, 5Z1.)

be correct representations of be-causal connections and, therefore, such connections have to be continuous as well.

In using this sloppy mode of speech, Leibniz was so preoccupied with the directionality of change as a by-product of causal or be-causal connections, that he ignored the fact that according to his system there cannot be a momentary state (either monadic or phenomenal) followed immediately by another one. There are of course passages in which he appears to be more careful. He does not use the idiom of a subsequent state following *its* preceding one for the purpose of talking about causal or be-causal connections. He speaks, instead, in the plural about our future thoughts or about our future states, specifying that they are contingent consequences of our preceding ones. This way of talking gets him off the hook, because it leaves room for continuity of change both monadic and phenomenal. The following passage from the *Discourse on Metaphysics* shows this quite clearly:

> Nothing can in fact happen to us except thoughts and perceptions, and all our future thoughts and perceptions are only the consequences, however contingent they may be, of our preceding ones, so that if I were capable of considering distinctly everything that is happening to me or appearing to me at this hour, I could see in it everything which will ever happen or appear to me. (L, 312: G, IV, 440.)

(ii) There are passages where Leibniz seems categorically to deny that monadic or even phenomenal change is genuinely continuous. Consider, for instance, the following short passage from a letter of Leibniz to De Volder, dated October II, 1705, "... changes ... are not truly continuous" (G, II, 279), or the following one from a letter to Electress Sophia of Hanover, dated October 31, 1705, in which, after having insisted that the duration of things is an infinite mass of momentary states, Leibniz emphatically asserts that, "... properly speaking..." there is no "...continuous passage from one state to the next one" (G, VII, 564.)

These passages considered in isolation can lead to the impression, that Leibniz thinks of change (monadic or phenomenal) as discretely and discontinuously structured. This is in one sense correct, because Leibniz conflates two notions in his account of continuity. According to him, a true or genuine continuum must be uninterrupted. Uninterruptedness as a property of true or genuine continua can be described by the following two conditions: (a) these continua are orders of possibles which contain no vacua; (b) they have no actual parts or ultimate constitutive units. Since actual change violates condition (b), it does not satisfy the property of uninterruptedness and therefore it is not continuous in this sense. But Leibniz also uses continuity in the sense of density, and in this sense change is continuous. Thus when Leibniz says that change is not "truly" continuous, he is indicating that change is not uninterrupted; but he is not denying that change is dense. That is, when Leibniz says that there is no continuous passage "from one state to the next one," he means that there is no *uninterrupted* passage, since to move from state to state involves passing through and therefore, as it were, stumbling upon intermediary ones. If asked, he would never have denied that change is continuous in the sense of being dense, as the following passage, from the same letter to De Volder (dated October 11, 1705) , indicates.

> Matter is not continuous but discrete, and actually infinitely divided, though no assignable part of space is without matter. But space, like time, is something not substantial, but ideal, and consists in possibilities, or in an order of coexistents that is in some way possible. And thus there are no divisions in it but such as are made by the mind, and the part is posterior to the whole. In real things, on the contrary, units are prior to the multitude, and multitudes exist only through units. (The same holds of changes, which are not truly continuous.) (R, 245; G, II; 278-279.)

In this passage, actual density of the spatially extended, and similarly of change, is equated with actual infinite divisibility. Additionally, genuine continuity is equated with the uninterruptedness and the priority of the whole over its parts of the genuine ideal continua.

(iii) The third reason for rejecting the discrete and discontinuous model as the correct interpretation of Leibniz's metaphysics of change has to do with his principle of continuity. If we were to admit it, continuity would be violated at the metaphysical level (i.e., at the real.) Additionally, we would be faced again with the problem of a world created by a God who likes perpetually to play deceiving games with us. We would be, in other words, falsely representing a discrete and discontinuous real change as a discrete and continuous one.

(iv) The fourth reason for not accepting the proposed model is not merely of a speculative or consequential, but of a positive and direct nature. There are, in other words, passages in which Leibniz unequivocally insists that monadic change is continuous, in the sense of being dense. Let us consider, for example, the following two passages the first from *The Monadology* and the second from *The Metaphysical Foundations of Mathematics* (written after 1714):

> 10. I also take it as agreed that every created being is subject to change, and therefore the created monad also, and further that this change is continuous in each one. (L, 643; G, VI, 608.)

> There is, moreover, a definite order in the transition of our perceptions when we pass from one to the other through intervening ones. This order, too, we can call a *path*. But since it can vary in infinite ways, we must necessarily conceive of one that is most simple, in which the order of proceeding through determinate intermediate states follows from the nature of the thing itself, that is, the intermediate stages are related in the simplest way to both extremes. (L, 671; GM, VII, 25.)

In the first passage we realize that Leibniz states quite clearly that change in each one of the created beings is continuous . Such a change could be thought of as taking place either at the methaphysical or the phenomenal level, since every monad moves from a monadic state A to a monadic state B representing distinctly, or confusedly, such a change in corresponding representational states of its own. From the second quoted passage, it becomes clear that such a change takes place from one representational or perceptual state to the next one, through intervening perceptual states, in the simplest possible way. That is, such a change or transition is not only continuous, in the sense of density, but also linear, the best model for simplicity of change being that of a straight line segment.

3. *The Discrete and Continuous Model*

According to this model both monadic and phenomenal change are discrete and continuous. A monad moves, at the metaphysical level, from state to state via an infinitude of densely arranged intermediate representational states following, as it were, a continuous, metaphysically temporal path. Phenomenal change is the correct representational counterpart of monadic change and, as such, it is the discrete and continuous phenomenal correlate of it. That is, a continuous metaphysical temporal path from a monadic (i.e., representational) state A to a monadic state B is represented as a continuous phenomenally temporal path from the representation A_1 of A, to the representation of B_1 of B. Appropriate inter-monadic correlations established by the pre-established harmony make it possible, by using the notion of a state of the real world of monads as an aggregate[45] of the metaphysically simultaneous representational states of all monads, to talk of simultaneity at both the metaphysical and the phenomenal levels.

Let us remind ourselves of the simple example of the two monads A and B that we used in order to explain simultaneity at both the metaphysical and phenomenal levels. We considered these monads to be in metaphysically

[45] Or, if we were to use appropriate mathematical terminology, as the equivalence class of the monadic states (of all monads) which are metaphysically simultaneous.

simultaneous representational states R_A and R_B, when A, by being in R_A, represents B as now being in R_B, and B, by being in state R_B, represents A as now being in R_A. Such a notion of simultaneity could be paired with a theory of time measurement which would have as its metaphysical basis temporal metric predicates as inherent characteristics of the representational structures of the monadic representors, that is, characteristics specifying the temporal distance between representations of different states of the real world of monads.

Before we give a formalized account of the interconnection between monadic and phenomenal change as it refers to density, we should address ourselves to the following question, which is somewhat related to the first two Zenonian paradoxes of motion. How can we accommodate the fact that in moving from a monadic state A to a monadic state B, a monad M passes through an infinitude of other states, given that a monad is a finite knower? Or how can we get over the difficulty that we represent an infinitude of representational states as temporally in between any two non-simultaneous states A and B, if we, as knowers, can never reach infinity? The answer to the question can be found in the following passage from *The Monadology*.

> ... a soul can read within itself only what it represents distinctly; it cannot all at once develop all that is enfolded within it, for this reaches to infinity. (L, 649; G, VI, 617.)

Although we represent the spatio-temporal world in its infinite entirety, we are able to represent distinctly, and therefore come to know, only finitely many of the representational items that co-exist in a total perspective representation. This answer, applied to the case of monadic change, can be transformed as follows: although we represent all states of the world of monads as they represent one another, we distinctly represent only finitely many parts of finitely many of them. The rest are represented confusedly, which could mean either that we are unconscious of them, or that we represent aggregates of them as single, undifferentiated, temporally extensive units. We could even adopt the extreme position that there are cases according to which a monad M moves from a state A to a state B passing through an infinitude of other states in such a way that each time it is conscious of being in the corresponding state. We have to supplement such a position, however, with the additional premise that the monad M does not have any epistemological access to such a fact. That is, the monad *does* not *know* that it went consciously through all those states because each time represents distinctly only finitely many items in a total representation and therefore it can only remember finitely many episodes of its previous life. A more plausible position to adopt would be the following:

Although the monad M moving from a monadic state A to a monadic state B passes through an infinitude of other states, it does that by being in a distinct representational state only finitely many times during such a journey. Additionally, the monad M, as a finite representor, can only remember finite parts of finitely many of its previous distinct representational states.

D. FORMALIZATION OF THE THIRD MODEL AS IT REFERS TO DENSITY OF MONADIC AND PHENOMENAL CHANGE

It is perhaps needless to say that we consider the third proposed model[46] as a correct interpretation of Leibniz's metaphysics of change. The basic reasons for holding this position are three:

There is definite textual support for it which coexists with the sloppy language Leibniz sometimes uses, especially when he is talking about causal or rather be-causal connections between a subsequent monadic (i.e., representational) state and *its* preceding one.

It conforms to the principle of continuity.

It is a correct explanatory model for the well-foundedness of phenomenal change which, according to Leibniz, is discretely and continuously structured.

In order to develop further such a model, so that we can bring into the picture the interconnection between monadic and phenomenal change as it refers to density, we need to adopt a first-order language with equality, "=", the vocabulary of which is as follows:

(1) The first group of symbols contains the usual connectives and quantifiers, plus a left and right parenthesis. It contains additionally an infinite countable number of variables "x_1", "x_2", "x_3", ... These variables will range over either monads or over real momentary states of the world of monads as they represent one another.

(2) The second group of symbols contains two subgroups, the cardinality of each of which is the same as the cardinality of the continuum. A typical symbol of the first group will be of the form "a_p", and a typical symbol of the second group will be of the form "b_r", where p and r are real numbers. The first group will provide us with names for monads, and the second with names for

[46] We called it the "discrete and continuous model" because we wanted to emphasize the fact that both monadic and phenomenal change can be thought of as *actual* linear continua, in the sense that in between any two *actual* (real or phenomenal) states A and B of the world there is always a third one which will play the role of the *present* before B and after A. "Actuality" has to be understood here as signifying the fact that each one of these states, each in its turn, *was, is, or will be* the discrete fabric of both the real and the phenomenal *present.*

real (i.e., representational) momentary states of the world of monads as they represent one another.

(3) The third group contains two unary predicate symbols "M" and , "S", a binary predicate symbol "M_B" and a ternary one "R_B". The symbol "M" when interpreted will single out monads and the symbol "S" real momentary states of the world of monads. The symbol "M_B" will apply to ordered pairs of real momentary states of the world of monads as they represent one another as representing. When this symbol is interpreted it will mean that the first member of the ordered pair *has been* or *is* or *will be* metaphysically present before the second. The symbol "R_B" will apply to ordered triples, the first member of which will be a monad, and the other two real momentary states of the world. When interpreted, the first member of the triple will represent the second member as phenomenally temporally prior to the third.

We can now see how the well-founded phenomenal relation of "temporally between" can be expressed in our formal language as a representation of what really obtains, i.e., as a representation of temporal betweeness at the level of metaphysical (real) change.

First, let us assume that b_1 and b_2 are real states of the world of monads as they represent one another. Then the fact that b_1 is metaphysically prior to b_2 can be expressed by:

$$M_B(b_1, b_2).$$

Second, if b_1, b_2, and b_3 are real distinct states of the world, the following three statements are true:

$$M_B(b_1, b_2) \vee M_B(b_2, b_1),$$
$$\neg M_B(b_1, b_1),$$
$$M_B(b_1, b_2) \wedge M_B(b_2, b_3) \rightarrow M_B(b_1, b_3).$$

Third, if a_1 is a monad and b_1, b_2, and b_3 are real states of the world and a_1 represents b_1 as phenomenally temporally prior to b_2 and b_2 as phenomenally temporally prior to b_3, then the following statements are true:

$$\neg R_B(a_1, b_1, b_1),$$
$$R_B(a_1, b_1, b_2),$$
$$R_B(a_1, b_2, b_3),$$
$$R_B(a_1, b_1, b_3) \wedge R_B(a_1, b_2, b_3) \rightarrow R_B(a_1, b_1, b_3).$$

Fourth, if b_1 is metaphysically temporally prior to b_2, any monad a_1 would correctly represent b_1 as phenomenally temporally prior to b_2. Conversely, if a_1 correctly represents b_1 as phenomenally temporally prior to b_2, then b_1 would be metaphysically prior to b_2. This fact could be expressed by the following statement:

$$M_B (b_1, b_2) \leftrightarrow R_B (a_1, b_1, b_2).$$

That a state of the world b_2 is truly metaphysically temporally between b_1 and b_3 can be expressed by the statement:

$$M_B(b_1, b_2) \wedge M_B(b_2, b_3).$$

Finally, a state of the world b_2 is phenomenally temporally between b_1 and b_3 if any monad a_p represents b_2 to be so. This fact can be expressed as:

$$R_B (a_p, b_1, b_2) \wedge R_B (a_p, b_2, b_3).$$

If we want to express density of phenomenal change as parasitic upon density of real change we must take into account two considerations: (i) Phenomenal change is dense with respect to every representor and (ii) under normal circumstances whenever a real state of the world is metaphysically temporally between two other such states, a representor represents it as phenomenally temporally between them and vice-versa.

Using (i) and (ii), we can express density of phenomenal change as parasitic upon density of real change using the following two axioms:

(1) $\forall x_1 \forall x_2 \forall x_3 \forall x_4 (M(x_1) \wedge S(x_2) \wedge S(x_3) \wedge S(x_4) \wedge x_2 \neq x_3 \wedge x_3 \neq x_4 \wedge x_2 \neq x_4 \rightarrow (M_B(x_2, x_3) \leftrightarrow R_B(x_1, x_2, x_3)) \wedge \neg M_B(x_2, x_2) \wedge (M_B(x_2, x_3) \vee M_B(x_3, x_2)) \wedge (M_B(x_2, x_3) \wedge M_B(x_3, x_4) \rightarrow M_B(x_2, x_4)))$,

(2) $\forall x_1 \forall x_2 (S(x_1) \wedge S(x_2) \wedge x_1 \neq x_2 \wedge M_B(x_1, x_2) \rightarrow \exists x_3 (S(x_3) \wedge x_1 \neq x_3 \wedge x_2 \neq x_3 \wedge M_B(x_1, x_3) \wedge M_B(x_3, x_2)))$.

Before we move on to examine Leibniz's solution to the problem of the composition of the continuum, some remarks are necessary concerning the third proposed model for interpreting the connection between monadic and phenomenal change, as well as its formalization, as it refers to what we would catachristically call temporal (phenomenal or real) density.

(i). Both phenomenal change and monadic change can be thought of as forming actual continua, i.e., as constituted by temporally unextended entities (namely, phenomenal or real instants) and second as satisfying density.

(ii). They are isomorphic, because, first, to every real instantaneous state of the world corresponds a phenomenal one and vice-versa; second, both are directional and linear in the same way; third, metric properties of the phenomenal change can be mapped in a coherent and consistent way onto isomorphic metric features of the real or monadic change.

We should point out that if we were to give an account of how we come to represent a *past* real state A of the world *now*, we would be using the following obvious model. We represent ourselves as *then* representing directly such a state, where the "now" and the "then" could be thought of as differentiated by their differing degrees of distinctness as representations in the framework of the same total representation. Such a difference could be thought of as coming from the fact that a phenomenal *present* is a temporally direct representation of what is metaphysically *present,* and a phenomenal *past* is a temporally indirect[47] representation of what was metaphysically *present.* That is, temporally indirect representations could and should be used to explain qualitative temporal differences between real states of the world as they are represented, i.e., differences pertaining to the rough non-quantitative distinction of *past*, *present*, and *future*. Yet this would not help much if we were to try to give a detailed explanatory account of the metric features of phenomenal change as based upon similar features of real (monadic) change. Two phenomenal states of the world A_l and B_l are separated by a phenomenal time distance t_P, because the real states of the world of monads A and B, which are represented as A_l and B_l, are separated by a metaphysical time distance t_m which can be thought of as having its predicative representational counterpart in the representational structure of every representor. Such a representational counterpart is intrinsically connected with the representations of A and B in the representor.

(iii). Concerning the directionality of both the monadic and phenomenal change, we have to take seriously Leibniz's contention that it is based upon a kind of metaphysical causality which is expressed as physical causality at the level of the phenomenal[48]. Appetition is the striving toward the future and

---

47 Where by the expression "temporally indirect representation" we mean one that has the basic characteristic of being the representation of a real state of the world B, inside the representation of a real state of the world A.

48 Additionally, at the level of the phenomena, we can be aware of such directionality because we can distinguish the past from the future. Although we represent them both, we are, on the one

"...every present state of a simple substance is a natural consequence of its preceding state, in such a way that the present is great with the future..." (*The Monadology,* L, 645; G, VI, 610.)

(iv). Simultaneity has to be cashed out in terms of the simple model of the two monads A and B that we proposed earlier. On this model simultaneity at the phenomenal level is based upon simultaneity at the level of the real (as dictated, of course, by pre-established harmony.) Such a position enables us to talk about a real momentary state of the world of monads as representing one another, and of its phenomenal counterpart as viewed perspectively and separately by each of the monadic representors. In one sense, from God's point of view, we live[49] in a non-perspective phenomenal world which we see perspectively.

(v). The linearity of both the monadic and the phenomenal change is beyond doubt for Leibniz. He thinks of them as the simplest continuous paths[50] for moving from one state to another through intermediate ones. The linearity of phenomenal change is, of course, parasitic upon the linearity of monadic change[51]. We formalized this contention using the statements

$$M_B(b_1, b_2) \vee M_B(b_2, b_1),$$
$$\neg M_B(b_1, b_1),$$
$$M_B(b_1, b_2) \wedge M_B(b_2, b_3) \rightarrow M_B(b_1, b_3),$$
$$M_B(b_1, b_2) \leftrightarrow R_B(a_1, b_1, b_2),$$

where a_1 is any monad and b_1, b_2, b_3 are real states of the world of monads as they represent one another. The last statement has, of course, a wider scope than just implying linearity of phenomenal change. It also spells out the truth that what is metaphysically "before" corresponds to what phenomenally appears as "before", and vice-versa.

(vi). The variables and the quantifiers we used in the formalization of metaphysical temporal density, phenomenal temporal density and their interconnection, range over either monads or states of the real world of monads as they represent one another. The mode of existence of monads is obviously different from that of their states or of states of the real world. With this in

hand, unconscious of our representations of the future and, on the other, we can distinctly remember finitely many instants or parts of the past.

49 That is, we come to think and act as if the world of phenomena was actually "out there."

50 See, again, GM, VII, 25.

51 To use Cartesian terminology, phenomenal change is change at the level of objective and monadic change at the level of formal reality.

mind, when we used the existential quantifier in talking about the existence of a real state of the world we meant that such a state *has been*, *is*, or *will be* the essential fabric that constitutes a specific present in the flow of real change or in the enfolding of what is temporally predetermined.

E. ACTUAL INFINITY AND THE LEIBNIZIAN SOLUTION TO THE PROBLEM OF THE COMPOSITION OF THE CONTINUUM

According to Leibniz, there are two labyrinths in which reason can go astray. The first concerns freedom and necessity, the second the composition of the continuum out of ultimate indivisible elements. Leibniz maintained that both have the same source, namely, the infinite. He also insisted that both have to be unraveled[52] before one could apply oneself to the task of establishing a solid metaphysics. In his early paper *On Freedom*, probably written around 1679, one can find the following passage:

> ... there are two labyrinths in which the human mind is caught. One concerns the composition of the continuum the other concerns the nature of freedom. And both arise from the same source, namely, the infinite. (L, 264; F, 180.)

Years later (*Theodicy*, 1710) he returns to the same theme adding that the first labyrinth perplexes everyone and the second "exercises philosophers only":

> There are two famous labyrinths where our reason very often goes astray: one concerns the great question of the Free and the Necessary, above all in the production and the origin of Evil; the other consists in the discussion of continuity and of the indivisibles which appear to be the elements thereof, and where the consideration of the infinite must enter in. The first perplexes

---

[52] How important Leibniz thought this task to be can be gathered from the following passages, which come from writings of his from different periods of his life:

> The whole labyrinth about the composition of the continuum must be unraveled as rigorously as possible. (L, 159; J, 34-36.)

> ... nothing but Geometry can furnish a thread for the labyrinth of the composition of the continuum, of maxima and minima, and of the unassignable and the infinite, and no one will arrive at a truly solid metaphysics who has not passed through that labyrinth. (R, 108-109; GM, VII, 326.)

> But if the knowledge of continuity is important for speculative enquiry, that of necessity is none the less so for practical application; and it, together with the questions therewith connected, to wit, the freedom of man and the justice of God, forms the subject of this treatise. (Th., pref., 54; G, VI, 29.)

almost all the human race, the other exercises philosophers only. (Th., pref., 53; G, VI, 29.)

Leibniz believed that he had provided the thread for finding the way out of the first labyrinth in his *Theodicy*. Leibniz's account of the way out of the second labyrinth, which we will consider exclusively in this section, was not similarly spelled out in a single work. This is not to say that we do not have enough information about the solution he proposed. On the contrary, numerous passages from his various writings can serve the purpose of establishing a unified picture by providing what Leibniz considered to be the final answer to the problem[53].

It was Leibniz's belief that the problem of the composition of the continuum (as well as the problem of free will) had its origin in the nature of the infinite. Questions pertaining to the existential status of the infinite, as well as to the distinction between the actual and the potential one, were considered by him as being immediately related to the problem. There is a sense in which Leibniz believed in an actual infinite. His belief was connected to his doctrine that the spatially extended was not only infinitely divisible, but actually infinitely divided[54]. Yet we have to be careful when we say that Leibniz was committed to the idea of the existence of the actual infinite. He definitely believed that to everything spatially extended there corresponded an infinitude of individual substances (monads) which, as represented, appear to constitute it. On the other hand, he denied the existence of a true infinite as belonging to the created world: "the true infinite exists, strictly speaking, only in the *Absolute*, which is anterior to all composition and is not formed by the addition of parts" (NE, 157; G, V, 144.) By denying the existence of a true infinite as

[53] It is important to notice that Leibniz did not think that the problems connected with the two labyrinths were entirely relevant to the everyday practice of a geometrician or a moral philosopher. He maintained that deductive procedures for carrying out demonstrations of theorems from given axioms and deliberations reached in accordance with a set of ethical premises could be more or less independent of these problems, the solutions of which are nonetheless important to philosophy and theology (and therefore to metaphysics.) The following passage from the *Discourse on Metaphysics* is illuminating in this respect:

> A geometrician does not need to encumber his mind with the famous labyrinth of the composition of the continuum, and no moral philosopher, and still less a jurisconsult or politician, needs to trouble himself with the great difficulties involved in reconciling free will with the providence of God, since the geometrician can carry through his demonstrations, and the politician finish his deliberations, without entering these discussions; yet they are nonetheless necessary and important in philosophy and theology. (L, 309; G, IV, 435.)

[54] See his reply to Foucher, *Journal des savans*, August 3, 1963 (W, 99; G, I, 416).

belonging to the created world of monads and of their representations, it seems that Leibniz wanted to stress the difference between the existence of infinite aggregates and of substantial infinite wholes[55], which could be considered by themselves as unitary. There is no true infinite in the created world because in it we cannot find a true substance consisting of an infinitude of simple substances. As Russell puts it, for Leibniz "an infinite aggregate is not truly a whole and therefore not truly infinite...[56]" Yet, Leibniz agrees that we have a positive idea of the infinite. Such an idea does not come as an inductive extension of the idea of what it is to be finite. In the *New Essays on Human Understanding* one can find the following very interesting passage:

> THEO. As to that, I would direct you to what I have said in several places in order to show that all these ideas, and especially that of God are within us from the outset that all we do is to come to pay heed to them; and that the idea of the infinite, above all, is not formed by extending finite ideas. (NE, 225; G, V, 209.)

The positive idea we have of the infinite is innate and anterior to that of the finite in the sense that it is the idea of a partless, unitary, unbounded immensity. It is a companion idea to the one we have of God. It is in that sense that it differs from the idea of an infinite aggregate and it does not constitute the ground for accepting the idea of an infinite number.

According to Leibniz, the idea of an infinite number is a faulty extension of the idea of a finite number. This is so for two reasons. First, there is no real substance consisting of an infinitude of simple substances and, therefore, there is nothing substantial and unitary for an infinite number to which it may refer. As one realizes from the following passage (*New Essays*) for Leibniz "there are an infinity of things" but there is no infinite number, or line, or infinite unitary whole:

> THEO. Properly speaking, it is true that there are an infinity of things, i.e., that there are always more of them than can be specified. But there is no infinite number, or line or any other infinite quantity, if these are understood as true wholes, as it is easy to prove (NE, 157; G, V, 144.)

Second, the idea of an infinite number is self-contradictory. In the following passage from a letter of his to Malebranche, dated June 22, 1679, Leibniz,

---

55 See, e.g., G, II, 304, (letter to Des Bosses, dated March 11, 1706) where he categorically asserts that an infinite aggregate is not one whole.

56 [128] p. 109.

equating the idea of an infinite number with that of the number of all numbers[57], gives us the argument which leads to the purported contradiction:

> Mons. Des Cartes in his reply to the *second objections*, article two, agrees to the analogy *between the most perfect Being and the greatest number*, denying that this number implies a contradiction. It is, however, easy to prove it. For the greatest number is the same as the number of all units. But the number of all units is the same as the number of all numbers (for any unit added to the previous ones always makes a new number.) But the number of all numbers implies a contradiction which I show thus: To any number there is a corresponding number equal to its double. Therefore the number of all numbers is not greater than the number of even numbers, i.e., the whole is not greater than its part. (R, 244; G, I, 338.)

It is rather unfortunate for Leibniz-and fortunate for us, epigones of Cantor, who can work using the conceptual richness of the Cantorian paradise of infinite numbers-that both arguments against infinite numbers are flawed. The first is defective because it can be used with equal force against the acceptance of finite numbers[58]. According to Leibniz there is no infinite number, if such a number is to be understood as referring to a substantial, collective, and at the same time unitary whole. But, similarly, there is no finite number (with the exception of number 1), if such a number is also to be understood as referring to a substantial, collective, and at the same time, unitary whole. Finite numbers, in other words share with infinite numbers their peculiar mode of existence. They can only refer to aggregates and not to real wholes. If we adopt the Leibnizian monadic metaphysics, a finite number can only be understood in the mode of a distributive and not of a collective whole. There is, for example, no real substance consisting of 12 monads; that is, to the number 12-and to any finite number, with the exception of 1-there is nothing substantially unitary to refer to.

Leibniz's second reason for not accepting infinite numbers is that the idea of such a number is self-contradictory. But the argument to this conclusion has as a premise the maxim that the whole is greater than any of its proper parts. If such a premise is not used, then the argument works perfectly well as a refutation of the maxim. If we adopt modern terminology, what the argument shows is that there is a one-to-one correspondence between the natural numbers and the even natural numbers. Furthermore, if we formally accept the

---

[57] Such an equating is not harmful to the argument, since Leibniz tacitly assumes that the numbers we gather together in order to produce the number of all numbers are not self-contradictory, i.e., they are finite numbers. A careful inspection of the passage reveals that they are special finite numbers, namely, natural numbers.

[58] Where by "finite numbers" we mean the natural numbers.

Cantorian hierarchy of infinites-with or without the ontology it presupposes-we can adopt the consistent view that proper subsets of infinite sets can have cardinality equal to that of the sets, since cardinality is defined via the notion of one-to-one correspondence. A *proper part/whole* relation, considered as a set-theoretic *proper subset/set* inclusion, never implies equality of the proper part to the whole, and the old maxim cherished by philosophers still holds. If, on the other hand, the *proper part/whole* relation is interpreted as indicating the nonexistence of a one-to-one correspondence between any proper infinite part of an infinite whole and the infinite whole itself, then the old maxim is violated.

The Leibnizian solution to the problem of the composition of the continuum is closely related to his ideas about the nature of the actuality of the infinite. The basic friction, as he saw it, was between the idea of the continuum as a unitary substantial whole and the indivisibles-namely, the substantial metaphysical points-which were supposed to constitute it. Such a friction had two parts. First, it was related to the question of how indivisible and unextended entities could form a continuously extended, unitary substantial whole, where by "continuously extended" we mean, among other things, "capable of continuous measurement." The predominant mathematical puzzle connected with this metaphysical question was related to the following variation of the *sorites* problem: how can a line segment consist exclusively of unextended units (mathematical points) so that extension becomes a property of a set of unextended entities, given that addition of a point to this line segment does not lead to change of its length[59]? Second, it was also related to the worry that reason would go astray if the indivisible constituents and the continuum, as a unitary substantial whole, could at the same time be thought of as being real[60]. That is, given a monadic, i.e., pluralistic, metaphysics, one wishes to know how mere infinite aggregates of real simple substances could form real, unitary, and continuous substantial wholes. For that matter, given a monistic metaphysics, it is unclear how a real, unitary, continuous whole could consist of an infinitude of real indivisible, unextended substantial units.

Leibniz used in his metaphysical argument concerning the denial of a real infinite-and especially of an infinite number-the doctrine that such an idea does not have a real substantial referent. On the same grounds he denied the existence of a real continuum. Yet he claims we have innately both the notion of an ideal continuum and of a positive true infinite. In the case of a positive

---

59 According to Leibniz, even an infinitude of points collected together would not make an extension. See, e.g., G, II, 370.

60 See Rescher [125], p. 104.

true infinite, the notion is parasitic upon our idea of God. Although Leibniz does not seem, explicitly at least, to maintain a similar view in the case of an ideal continuum[61], the similarities between the two notions are strikingly apparent. Both are prior to their parts, in the sense that such parts are not actually, but only potentially there, and both have homogeneity and uninterruptedness as their characteristics. They are not amenable to any discrete enumeration, i.e., there is no infinite number corresponding to them indicating the cardinality of their basic constitutive parts, because such parts or units are not actually there. Additionally, in his final solution to the problem of the composition of the continuum, Leibniz used a contrast between the notion of an ideal continuum and that of an infinite aggregate of monads (which he thought not to be constitutive of it), that runs parallel to the contrast between the notion of the positive true infinite and that of an actual infinite as referring to an actual infinite aggregate of monads (which does not lead to the acceptance of an infinite number.) In order to see what Leibniz's solution is to the problem of the composition of the continuum, we have to examine it in relation to his tri-partite metaphysics, i.e., in relation to the ideal, the phenomenal, and the real levels of his system.

a. *The ideal*

According to Leibniz, reason becomes trapped in the labyrinth of the continuum because we tend to confuse the ideal and the actual. The only continua deserving to be so called are ideal. They are characterized by their lack of constitutive parts. They are prior to them in the sense that parts are the potential resultants of possible acts of division. Parts, in other words, are indeterminate, in that they are neither actual nor definite before a particular act of division is carried out. Additionally, there are no privileged positions for creating a partition through a mental act of division, as would be the case if there was a set of positions that was not dense, at which positions alone acts of division could be performed.

Leibniz in a passage from a letter of his to De Volder, dated January 19, 1706, states quite clearly that:

> ... in actual bodies there is only a discrete quantity, that is, a multitude of monads or of simple substances, though in any sensible aggregate or one corresponding to the phenomena, this may be greater than any given number.

[61] One reason for this could be that he hesitated to express such a view because he had to deal with more than one notion of ideal continua (as for instance those of space and time.) He would therefore have to explain their differences by referring to God's corresponding attributes. Such a move would place him on very slippery explanatory terrain.

> But a continuous quantity is something ideal which pertains to possibles and to actuals-in virtue of their being possibles as well. A continuum, that is, involves indeterminate parts, while, on the other hand, there is nothing indefinite in actual things, in which every division is made that can be made. Actual things are composed as a number is composed of unities, ideal things as a number is composed of fractions; the parts are actual in the real whole, but not in the ideal whole. But we confuse ideal with real substances when we seek for actual parts in the order of possibles and indeterminate parts in the aggregate of actuals and so entangle ourselves in the labyrinth of the continuum and in contradictions that cannot be explained. (L, 539; G, II, 282.)

By "parts" Leibniz does not mean points. Points are not constitutive parts of an ideal continuum, whether or not such parts are indeterminate. They are extremities[62] or potential boundaries[63] which get actualized by a certain mental act of division. Such a mental act transforms the potentiality of an indeterminate partition of the ideal continuum into actuality. Points can also be thought of as limits of an ideal infinite process, i.e., they can be thought of as "mere fictions generated by an infinite extension of a necessarily finite process[64]." In short, points can be thought of as extremities or potential boundaries, or limits and not as parts of a continuum, because they obey the following general maxim: parts of an n-th dimensional continuum can only be nth-dimensional chunks of it. On the other hand, m-th dimensional continua (points, lines, planes, etc.) in the framework of an n-th dimensional continuum (where m is greater than or equal to 0, n greater than or equal to 1, and n>m) can only be thought of as extremities or limits of an ideal infinite process.

The most important characteristics which, Leibniz thinks, belong to an ideal continuum (with the continua of space and time as our paradigm cases) are the following five:

1. Every such continuum belongs to the realm of innate mathematical ideas.

2. An ideal continuum is not composed of parts. Its parts are in a quite specific sense indeterminate, i.e., they are the potential resultants of possible, future mental acts of division which can be carried out anywhere throughout such a continuum.

3. An ideal continuum is not composed out of points. Moreover, an n-th dimensional continuum is not composed of m-th dimensional entities, where m

---

62 See, e.g., G, IV, 478; G, III, 622.

63 Especially if the continuum is linear. More generally, in the case of an n-th-dimensional continuum, n-1th-dimensional continua, where n is greater than or equal to 2, play the primary role of such potential boundaries.

64 See, [125], p. 103.

is greater than or equal to 0, n is greater than or equal to 1, and n>m. This is a generalization of the Leibnizian way out of what Rescher calls the "mathematical wing of the labyrinth of the continuum[65]."

4. An ideal continuum is everywhere dense. For instance, in the case of a straight line, between any two potential positions for carrying out corresponding mental acts of division, there is a third potential position for another such act to be performed.

5. Every ideal continuum is characterized by homogeneity[66]and uninterrupted- ness.

Homogeneity is the result of the fact that such a continuum is a kind of mental grid with undifferentiated and uniform potential parts, abstractly representing a certain continuous order of possibles. Space, for instance, is homogeneous because it is the continuous order of possible coexistents. A position A in space, as indicating a possible monadic point of view, cannot be differentiated from a position B, without a preexisting referential system which would thereby break down the homogeneity by making possible the assignment of different absolute coordinates to A and B. Homogeneity also breaks down at the level of the phenomena, because actualities (i.e., monads representing themselves and represented by others as occupying spatial positions) involve a determination within possibilities, which is characterized by particularities of what is actual. That is, once a monad M is represented as occupying a position A (with respect to a set of other monads), that position is uniquely

---

[65] According to Rescher:

> Leibniz emerges from the mathematical wing of the labyrinth of the continuum by dismissing the problem-task of building a line up out of points ([125], p. 103.)

[66] As we have already said, with regard to homogeneity, one can think of the possible exception of time, in the case of which we have a directional and therefore non-homogenous linear continuum. In a letter of his to Louis Bourguet, dated August 5, 1715, Leibniz states:

> I admit, however, that there is this difference between instants and points - one point of the universe has no advantage of priority over another, while a preceding instant always has the advantage of priority, not merely in time but in nature, over following instants. (L, 664; G, III, 581-582.)

On the other hand, according to the following passage from the *Fifth Paper to Clarke*, if we examine the structure of the potential parts of time, without appeal to its directionality as a whole, we can insist that even time is homogeneous.

> 27. The parts of time or place, considered in themselves, are ideal things; and therefore they perfectly resemble one another like two abstract units. But it is not so with two concrete ones, or with two real times, or two spaces filled up, that is, truly actual. (A 63; G, VII, 395.)

characterized and A is therefore completely differentiated from all the other positions, which are also uniquely characterized as being phenomenally occupied by monads distinct from M. Uninterruptedness, as a characteristic of an ideal continuum, can be thought of as indicating two things, first, the non-existence of vacua in the specific order of possibles that such a continuum is supposed to be, and second, the non-existence of actual constitutive units or parts upon which one would stumble, so to speak, if one were somehow to travel through such a continuum.

b. *The phenomenal and the real*

Leibniz maintains that the only genuine continua are the ideal ones. If we were to single out the most basic characteristics of such continua we would say that the first concerns their potential infinite divisibility, the second their uninterruptedness as indicating (i) the non-existence of vacua in the orders of possibles that such continua are supposed to be, and (ii) the non-existence in them of actual parts or constitutive ultimate units. Leibniz maintains that we cannot find genuine continua at the level either of the real or of the phenomenal. According to Leibniz (letter to Nicolas Remond, dated, March 14, 1714):

> The source of our difficulties with the composition of the continuum comes from the fact that we think of matter and space as substances, whereas in themselves material things are merely well-regulated phenomena, and *space is exactly the same as the order of coexistence, as time is the order of existence which is not simultaneous*. Insofar as they are not designated in extension by factual phenomena, parts consist only in possibility; there are no parts in a line except as there are fractions in unity. But if we assume that all possible points actually exist in the whole-as we should have to say if this whole were a substantial thing composed of all its parts-we should be lost in an inextricable labyrinth. (L, 656; G, III, 612.)

The above quoted passage is illuminating because it contains precisely and succinctly Leibniz's rationale for denying the existence of a real genuine continuum. The only real substances are the indivisible, unextended monads. There is no real genuine continuum, because if there were, it would not be a unitary whole; it would consist, that is, of an infinitude of monads and, therefore, it would not be a real being for it would not be truly one being[67]. It

[67] We should remind ourselves that a similar metaphysical argument was used by Leibniz against the idea of an infinite number. That argument held that the idea of an infinite number is incoherent because there is nothing substantial and unitary as its referent. The existence of infinite aggregates was considered by him as simply appropriating the acceptance of the actual infinite as a distributive and not as a collective notion.

is, after all, a Leibnizian doctrine that "what is not truly one being is also not truly a *being*" (G, II, 97.) Additionally, if there were a real genuine continuum, it would be an extended continuous whole composed exclusively of unextended substantial entities. But Leibniz maintains that even an infinitude of unextended entities collected together would not make an extension[68].

We would not have examined the Leibnizian solution to the problem of the composition of the continuum in all of its ramifications, if we were to ignore the following question. Given that there are no genuine continua at the level of the real, is it at least true that we can find such entities in the realm of well-founded phenomena? Leibniz's answer, as the last quoted passage testifies, is again negative. An extended material object is a phenomenon *bene fundatum*. As such, it is a representation of an infinite multitude of monads as representing one another in a coherent way. This extended material object, and for that matter the whole physical universe, satisfies, in a specifically modified sense[69], at least one of the basic criteria for being called a continuum. It is *actually* everywhere dense. That is, between any two monads A and B represented as being spatially positioned, there is a third monad C also represented as being spatially positioned. Nonetheless, the other important criterion for qualifying as a genuine continuum is not satisfied. Uninterruptedness, not as indicating the nonexistence of vacua, but as indicating the non-existence of actual parts or of constitutive ultimate units, is violated. The physical universe, at any particular moment, is a physical plenum in the sense that (a) it is the representation of the order of actual coexistents and not the representation of a unitary substantial whole, and (b) as such it consists of an infinitude of representations (one for each monad) so interwoven that the total representation resulting from it is characterized by actual density. That is to say, such a total representation, according to Leibniz, does not qualify as a genuine continuum for the following two reasons. First, it is not by itself a substantial, unitary, and potentially dense whole. Second, by being a representation of a discretely structured and metaphysically atomic reality, it is necessarily also discretely structured and representationally atomic. It is needless to say that this attitude is not restricted to the phenomenality of the spatially extended, but also extends to phenomenal change, as well as to every other aspect of the discrete world of monads (or their states) as they represent all the others as representing.

---

[68] See, again, G, II, 370.

[69] In that density in that case is a property referring to the order of the actuals and not to the order of the possibles.

In short, the Leibnizian solution to the problem of the composition of the continuum had two components. First, the ideal continua are not composed of constituent parts or constituent ultimate units. Second, at the levels of the real or the phenomenal the ultimate substantial units or their representations cannot and do not compose genuine continua.

The Leibnizian solution is heavily loaded with ontological considerations and worries. The main problem that Leibniz was trying to solve was two-fold and it involved the following two questions: (a) how we can build up a substantial, unitary, continuous whole out of indivisible, unextended, ultimate substantial units, and (b) how we can make sense of an ontology of basic unextended entities as giving rise to extended continuous aggregates. In the course of struggling with these questions, Leibniz produced a powerful metaphysical system, the enormous potential of which he could not and did not entirely realize. His solution to the problem of the composition of the continuum was impressive, but, in a sense, half-baked. This is not simply an anachronistic contention; we do not intend to compare his solution in a crude and perhaps unfair way with modern ones. The crucial and more important point is that there was enough room in his metaphysics for a finer solution, one with remarkable conceptual proximity to sophisticated modern ideas about the composition of the continuum.

There are two points we want to emphasize here. The first concerns, again, uninterruptedness as a property of a genuine continuum. The second concerns the variation of the old sorites problem that we mentioned earlier. Leibniz never quite managed to solve it satisfactorily, despite the fact that his metaphysics would have allowed him to do so.

Considerations involving the idea of a continuously extended whole and the nonexistence of a unitary substantial entity in the created world corresponding to it led him to believe, first, that genuine continua can only be ideal entities and, second, that, such continua can be characterized by potential infinite divisibility and uninterruptedness. It was the lack of uninterruptedness, as resulting from a pluralistic monadic metaphysics of substance, that led him to deny the existence of real or even of phenomenal genuine continua. He relied so heavily on this property that he restricted the term "continuum" to ideal ones. He sometimes went even further, calling "continuous quantities" only those ideal ones which pertain to the various orders of the possible and not of the actual (G, II, 282), as if density or infinite divisibilty were not that important in the characterization of a quantity as continuous. He therefore appears to be terminologically inconsistent, given that there are places where he uses such terms as "continuum," "physical continuum," or "continuous" as indicating basically density and not uninterruptedness. Consider, for example, the following passage from a letter of Leibniz to Des Bosses, dated May 29, 1716:

> There is continuous extension whenever points are assumed to be so situated that there are no two between which there is not an intermediate point. (R, 247; G, II, 515.)

It is more than clear that in this passage Leibniz considers density as the basic characteristic of "continuous extension". In the following passage, from a letter again to Des Bosses, dated January 24, 1713, he insists that unboundedness is a characteristic of "the physical continuum". There is no direct mention of density but it should be considered as being implicitly included as the fundamental characteristic of "the physical continuum", since not "stopping anywhere" would not obtain if density were violated. On the other hand, uninterruptedness, at least in the sense of non-existence of actual parts or constitutive ultimate units, is not a characteristic of the physical continuum and it is therefore excluded from the picture given.

> In the hypothesis of mere monads, the infinitude of the physical continuum would depend not so much on the principle of the best as on the principle of sufficient reason, because there is no reason for limiting, or ending or stopping anywhere. (G, II, 475).

Furthermore, Leibniz's principle of continuity, as applying to every level of his metaphysics, shows-although he is not fully aware of this-that the most important feature of continuity was density and not uninterruptedness[70]. The maxim that "nature never makes leaps," as applying to, e.g., real or monadic change (which according to Leibniz is in one sense discretely structured (G, II, 278)), dictates that such a change is continuous. That is, though uninterruptedness, as indicating the non-existence of ultimate unextended and indivisible units[71] is violated in this case, monadic change is termed continuous because it is characterized by density. Consider, again, the following passage from *The Monadology:*

---

[70] There are even places where the picture that Leibniz presents is totally confusing, as when he insists that:

> The continuum, however, though it has such indivisibles everywhere is not composed of them ... (R, 247; G, I, 416.)

without specifying what it means to say that the continuum "has such indivisibles."

[71] Such units are, in this case, the momentary monadic states.

> 10. I also take it as agreed that every created being is subject to change and therefore the created monad also, and further that this change is continuous in each one. (L, 643; G, VI, 608.)

The conclusion we are driving at is that continuity, as indicating density, is a characteristic of every level of the Leibnizian metaphysics. In the case of monadic reality, monads considered not as empty substantial shells, but as carriers of their personal, individualized, representational structures form not real wholes, but real aggregates which are everywhere dense. For instance, to the density of the spatially extended corresponds real density as specified by the model presented in III.B, or as specified by any model isomorphic to it. A similar point obtains in the case of monadic change. At the level of well-founded phenomena, density can be thought of as the representational isomorphic feature of the density that obtains at the level of the real. Both levels are characterized by *actual* density. That is, they are characterized by the property according to which, between any two actual entities (real or representational), there always exists an actual (real or representational) intermediary one.

Continuous structures at these levels are actual dense aggregates of indivisible units. Such actual dense aggregates are not substantial unitary wholes and therefore cannot be thought of as satisfying the property of uninterruptedness[72]. This is the trap Leibniz fell into. He considered uninterruptedness as the most important property of the continua and accordingly relegated actual continua to the level of non-genuine ones. This property is a sufficient, but not a necessary condition for continuity. It is a property that monistic theories of substance wrongly imposed upon it, during centuries of antagonism with pluralistic ones. It is also a property of Aristotelian origin. For Aristotle the line is uninterrupted in the sense that it is prior to its points, which come to be only through specific acts of division of the line.

The combination of a discrete and at the same time dense constitution of actual continuous structures was quite disturbing to Leibniz, who most of the time could not get over the idea of an aggregate as an unstructured pile of units. In other words, although Leibniz had the means to tackle the problem of

[72] There is of course another sense of uninterruptedness which we should not confuse with the one we gave. It refers to the impression we get of uninterruptedeness when we perceive, for example, a homogeneously colored surface. Such uninterruptedness is illusory and is due to characteristics of our perceptions such as sameness of the degree of distinctness or confusedness of our representations of the monads constitutive of the perceived surface. From God's, or the metaphysician's, point of view actual divisibility of what we perceive as undifferentiated is an undeniable ontological fact behind the deficiencies of our representational structure.

the composition of the continuum positively and not just negatively, he never realized this, because he was preoccupied on the one hand with the idea of the non-existence of a created substantial, continuous, unitary whole and, on the other, with the idea of the discreteness of the actual as opposed to the desired smooth homogeneity of a genuine continuum. The ideal level was for Leibniz the only level where he could find such continua. These ideal entities had the right properties; they were potentially dense and at the same time smooth, undifferentiated and uninterrupted unitary wholes.

If Leibniz had dropped the uninterruptedness condition-and he could have done this without sacrificing his monadic metaphysics, or his belief that there is no real or phenomenal continuous unitary entity-he would have come remarkably near to a modern view of what a continuum is[73] and how it is composed. By keeping the uninterruptedness condition, he thought he had solved the problem by denying that actual continuous aggregates are genuine continua and by insisting that the ideal continua, which are genuine, are not composed of indivisible unextended units. We think he did not solve the problem of the composition of the continuum, but simply drew, so to speak, a definitional line between genuine and non-genuine continua, or-to be terminologically closer to Leibniz-between things which do and things which do not qualify as continua. What he did was to define the notion of the continuum so that only ideal ones could qualify, as such. After this, his task was easy. There is no problem of "the composition of the continuum" because, by definition, continua are not composed of parts or of ultimate, indivisible, unextended units.

What would be an alternative, positive Leibnizian solution to the problem? Let us assume that density was the predominant feature of the continuous for him. If we once more consider space, the spatially extended, and its metaphysical correlate, we would say that at the level of the real there is an infinitude of monads which, at any particular moment form, together with their representational structure, a real continuum that is isomorphically represented as a spatial continuous actual plenum. Such a representation belongs to the level of the phenomenal and qualifies as a continuum because density is one of its characteristics. On the other hand, at the level of the ideal we have an entity which can play the role of a three-dimensional, continuous, and uninterrupted plenum, which could be considered as the abstract representation of the spatial,

---

[73] With the obvious qualification that a basic ingredient was missing from his account, namely, that of sequential completeness. In II.C we argued that he in some sense had this ingredient in his hands also, although he never completely realized it.

continuous, actual plenum; that is, its structural equivalent when all particularities of what is actual are removed.

Consider now how near such a metaphysical scheme is to the following modern picture. Let us assume that we can draw or imagine that we draw a *straight line* on an actual infinite blackboard. Such a line, once drawn, could be thought of as the equivalent of a Leibnizian phenomenal linear continuum. Move now to the Cantorian set-theoretic paradise and consider an *infinite set of points A* (of cardinality equal to that of the real line) and a *set of positional relations* or, better, a *set of positional predicates* P, related to the *set of points A*. Let us assume further that if we were isomorphically to represent the structure (A,P) on our blackboard, we would do so by drawing the straight line we drew before. We can imagine the structure (A,P) as the Leibnizian equivalent of an infinite multitude of monads being in completely specified representational states at a particular moment. Every such representational state, as referring to a corresponding particular monad, would be thought of as containing the equivalent of all positional predicates that specify the phenomenal position of the monad. Finally, let us imagine an *ideal straight line* (with points only potentially present), uninterrupted and homogeneous. The above three continua (that is the structure (A, P), the straight line we drew on the blackboard and the ideal straight line we imagined) are isomorphic in the sense that there are functions from any one of them to any one of the others, which are bijective and respect all the positional characteristics of the elements of these continua. As a byproduct of this, to any real point α that belongs to A corresponds a unique phenomenal point on the line we have drawn, and vice-versa; and to any such phenomenal point corresponds a unique potential point[74] in the ideal line and vice-versa.

If Leibniz had adopted this position he would not have had to worry about the variation of the sorites problem that we mentioned earlier either. The sorites problem is produced if we think of a set of points devoid of any positional structure. Leibniz makes exactly that mistake when he insists that even an infinitude of points collected together would not make an extension (G, II, 370.) An infinitude of points *can* make an extension if they are considered as already possessing a certain positional structure. To put it metaphysically, at any particular moment, monads (or metaphysical points) possess such a spatially positional structure[75], each one having its own, so that

[74] Which can be thought of as such because of the precondition of uninterruptedness.

[75] Monads possess a representational structure, a specific part of which, concerning spatial representations, we could call their spatial positional structure.

all together form a continuous[76] real aggregate which does not have to be substantially unitary. Moreover, a monad in such a situation represents itself as being positioned with respect to all the others and therefore as being at a specific distance from each one of them. The specificity of any such distance is not a function of the number of monads which are represented as spatially in between. As it has been mentioned, in a linear continuum the cardinality of a line segment is equal to the cardinality of any other line segment. The specificity of the distance between two monads A and B, as represented, is simply dependent upon the representational structure of the particular monads involved and has nothing to do with the number of monads that are represented as spatially between them. There is a sense in which the spatial distance between two monads A and B, as represented, depends upon indirect representation and therefore upon the set of monads represented as spatially positioned between A and B. Consider, for example, the following passage from *The Metaphysical Foundations of Mathematics* (written after 1914):

> In either order (of space or of time) [points] are considered nearer or more remote, according as, for the order of comprehension between them, more or fewer are required. (R, 247; GM, VII, 18.)

As it has been mentioned, if one wants to make sense of this, one has to employ the *proper subset/set* inclusion relation (and not the cardinality comparison) as follows: if, for example, there were three monads A, B, and C represented as co-linear and B is representationally between A and C, then the set of monads represented as spatially between A and B and the set of monads represented as spatially between B and C are proper subsets of the set of monads represented as spatially between A and C. Such a fact, although in accordance with metric considerations, is not a substitute for them.

[76] At any particular moment, representationally between any two monads A and B there is a third monad C. This representational fact as represented is what we call spatial density at the phenomenal level.

BIBLIOGRAPHY

[1] Adam, Ch. and Tannery P. (eds.) *Oevres de Descartes.* Vrin, J. /C.N.R.S., Paris, 1964-1976. This is a revised version of the Cerf edition, published in 12 volumes, 1897-1910.

[2] Alexander, H.G. (ed.) *The Leibniz-Clarke Correspondence.* Manchester University Press, Manchester,1970.

[3] Allison, H. E.(ed. and trans.) *The Kant-Eberhard Controversy.* The John Hopkins University Press, Baltimore and London, 1973.

[4] Anapolitanos, D. A. "Leibniz on Density and Sequential or Cauchy Completeness". In P. Nicolacopoulos (ed.) *Greek Studies in the Philosophy and History of Science,* Boston Studies in the Philosophy of Science. Kluwer Acad. Publ., vol. 121, pp. 361-372, 1990.

[5] Anderson, J. L. *Principles of Relativity Physics.* Academic Press, 1967

[6] Apostol, T. M. *Mathematical Analysis: A modern Aproach to Advanced Calculus.* Addison-Wesley Student Series Edition,Reading Massachusettes, Fifth Printing, 1971.

[7] Ariew , R. and Garber, D. (eds. and trans.) *Leibniz: Philosophical Essays.* Hackett Publ. Co., Indianapolis, Cambridge Mass., 1989.

[8] Aristotle. *Metaphysics.* Jaeger , W. (ed.) Oxford University Press, 0xford,1957.

[9] Aristotle. *Physics.* Ross, W.D. (ed.). Oxford University Press, 0xford,1973.

[10] Barbour, J. B. "Relational Concepts of Space and Time". *British Journal for the Philosophy of Science,* vol. 33, pp. 251-274, 1982.

[11] Barnes, J. (ed.) *The Complete Works of Aristotle.* Princeton University Press , Princeton, 1984.

[12] Brandom, R.B. "Leibniz and Degrees of Perception". *Journal of the History of Philosophy,* vol. XIX, pp. 447-479, 1981.

[13] Broad, C. D. "Leibniz's Last Controversy with the Newtonians". *Theoria* , vol. 12, pp. 143-168, 1946.

[14] Broad, C.D. "Leibniz's Predicate-in-Notion Principle and Some of its Alleged Consequences". In H.G. Frankfurt (ed.) *Leibniz: A collection of Critical Essays.* University of Notre Dame Press, Notre Dame, 1972, pp. 1-18.

[15] Broad, C.D. *Leibniz: An Introduction.* Cambridge University Press, Cambridge, 1975.

[16] Brown, G. "Compossibility, Harmony and Perfection in Leibniz". *The Philosophical Review,* vol. XCVI, pp. 173-203, 1987.

[17] Buchdahl, G. *Metaphysics* and *the Philosophy of Science. The Classical Origins: Descartes to Kant.* Basil Blackwell, Oxford, 1969.

[18] Burkhardt, H. and Degen, W. "Mereology in Leibniz's Logic and Philosophy". *Topoi,* vol. 9, pp. 3-13, 1990.

[19] Buchenau, A. (trans.) and Cassirer, E. (ed.) *G.W. Leibniz: Philosophische Werke.* In 2 volumes. Meiner, F. Leipzig, 1924.

[20] Cantor, G. *Contributions to the Founding of the Theory of Transfinite Numbers.* Dover, New York, 1955.

[21] Cassirer, E. *Leibniz System in seinen wissenschaftlichen Grundlagen.* N.G. Elwert, Marburg an der Lahn, 1902.

[22] Chang, C. C. and Keisler, H. J. *Model Theory.* North Holland, Amsterdam, 1973.

[23] Cottincham, J. , Stoothoff, R . and Murdoch, D. , (trans.) *The Philosophical Writings of Descartes.* Cambridge University Press, Cambridge, 1984.

[24] Couturat , L. (ed.) *Opuscules et fragments inédits de Leibniz.* In 2 volumes. Felix Alcan, Paris, 1903.

[25] Couturat, L, *La Logique de Leibniz d'après documents inédits.* Felix Alcan, Paris, 1901.

[26] Couturat, L. "On Leibniz's Metaphysics". In H.G. Frankfurt (ed.) *Leibniz: A Collection of Critical Essays.* University of Notre Dame Press, Notre Dame , pp. 19-46, 1972.

[27] Duncan, G.M. (trans.) *The Philosophical* Works *of Leibniz.* Second edition. Tuttle, Morehouse and Taylor Company, New Haven, 1908.

[28] Earman, J. "Who's Afraid of Absolute Space". *Australasian Journal of Philosophy,* vol. 48, pp. 287-317, 1970.

[29] Earman, J. "Infinities, Infinitesimals and Indivisibles: The Leibnizian Labyrinth". *Studia Leibnitiana* , vol. VII, pp. 236-251, 1975.

[30] Earman, J. "Perceptions and Relations in the Monadology". *Studia Leibnitiana* , vol. IX, pp. 212-230, 1977.

[31] Earman, J. *World Enough and Space-Time. Absolute Versus Relational Theories of Space and Time.* The MIT Press, Cambridge, Massachusetts,1989.

[32] Einstein, A. "Die formale Grundlage der allgemeinen Relativitätstheorie". *Preussische Akademie der Wissenschaften* , Berlin, Sitzungsberichte, pp. 831-839, 1914.

[33] Einstein, A. "On The Electrodynamics of Moving Bodies", 1905 Translated and reprinted in W. Perret and G. B. Jeffrey (ed.) *The Principle of Relativity.* Dover, New York, pp. 37-65, 1952.

[34] Einstein, A. "The Foundations of the General Theory of Relativity", 1916. Translated and reprinted in W. Perret and G.B. Jeffrey (ed.) *The Principle of Relativity.* Dover, New York, pp. 111-173, 1952.

[35] Einstein, A. *The Meaning of Relativity.* 5th edition. Princeton University Press, Princeton, 1955.

[36] Einstein, A. *Relativity: The Special and the General Theory.* Bonanza Books, New York, 1961.

[37] Einstein, A. and Grossmann, M. "Entwurf einer verallgemeinerten Relativitätstheorie und einer Theorie der Gravitation". *Zeitschrift für Mathematik und Physik,* vol. 62, pp. 225-261, 1913.

[38] Einstein, A. and Grossmann, M. "Kovarianzeigenschaften der Feldgleichungen der auf die verallgemeinerte Relativitätstheory gegrundenten Gravitationstheorie". *Zeitschrift für Mathematic und Physik,* vol 63, pp. 215-225,1914.

[39] Erdmann, J.E. (ed.) *God. Guil. Leibnitii Opera Philosophica quae extant Latina, Gallica, Germanica Omnia.* In 2 volumes, S.G. Eichleri, Berlin 1840.

[40] Erlichson, H. "The Leibniz-Clarke Controversy: Absolute Versus Relative Space and Time". *American Journal of Physics,* vol. 35, pp. 89-98, 1967.

[41] Enderton, H. B. *A Mathematical Introduction to Logic.* Academic Press, New York, 1977.

[42] Enderton, H. B. *Elements of Set Theory.* Academic Press, New York, 1977.

[43] Farrer , A. (ed.) and Huggart, E.M. , (trans.) *G.W. Leibniz: Theodicy.* Open Court, La Salle, 1985.

[44] Fleming, N. "On Leibniz on Subject and Substance". *The Philosophical Review,* vol. XCVI, pp. 69-95, 1987.

[45] Foucher, A. de Careil (ed.) *Nouvelles lettres et opuscules inédits de Leibniz.* A. Durand, Paris, 1857.

[46] Fox, M. "Leibniz's Metaphysics of Space and Time". *Studia Leibnitiana* , vol. II, pp.29-55, 1970.

[47] Fraenkel, A.A. *Abstract Set Theory.* North Holland, Amsterdam, 1968.

[48] Fraenkel, A. A. , Bar-Hillel , Y. and Levy, A. *Foundations of Set Theory.* North Holland, Amsterdam, 1973.

[49] Frankfurt, H.G. (ed.) *Leibniz: A Collection of Critical Essays.* University of Notre Dame Press, Notre Dame, 1972.

[50] Friedman, M. *Foundations of Space-Time Theories.* Princeton University Press, Princeton, 1973.

[51] Friedman, M. "Kant's Theory of Geometry". *The Philosophical Review,* vol. XCIV, pp. 455-506, 1985

[52] Furth, M. "Monadology". In H.G. Frankfurt (ed.) *Leibniz: A Collection of Critical* Essays, University of Notre Dame Press, Notre Dame, pp. 99-136, 1972.

[53] Gale, G. "The Physical Theory of Leibniz". *Studia Leibnitiana* , vol. II, pp.114-126, 1970.

[54] Gale, R. M. (ed.) *The Philosophy of Time: A Collection of Essays.* Humanities Press, New Jersey, 1968.

[55] Garber, D. "Leibniz and the Foundations of Physics: The Middle Years". In K. Okruhlik and J. B. Brown (eds.) *The Natural Philosophy of Leibniz,* D. Reidel, Dordrecht, pp. 27-130, 1985.

[56] Gavroglu, K. "Theoretical Frameworks for Theories of Gravitation: A Case of a «Sui Generis» Research Program". *Methodology and Science,* vol. 19, pp. 91-123, 1986.

[57] Gebhardt, C. (ed.) *Spinoza Opera.* In 4 volumes. Carl Winter, Heidelberg, 1925.

[58] Gerhardt, C. I. (ed.) *Leibnizens mathematische Schriften.* In 7 volumes. Volumes 1-2 published by A. Asher and Co., Berlin, volumes 3-7 published by H.W. Schmidt, Halle, 1849-1863.

[59] Gerhardt, C. I. *Die philosophischen Schriften von Gottfried Wilhelm Leibniz.* In 7 volumes. Weidman, Berlin, 1875-1890.

[60] Grene, M. and Ravetz, J. R. "Leibniz's Cosmic Equation: A Reconstruction". *The Journal of Philosophy,* vol. LIX, pp. 141-146, 1962.

[61] Grua, G. (ed.) *G.W. Leibniz: Textes inédits d'après les Manuscripts de la Bibliothèque Provincial de Hanovre.* Presses Universitaire de France, Paris, 1948.

[62] Grunbaum, A. *Philosophical Problems of Space* and *Time.* D. Reidel, Dordrecht, 1973.

[63] Gueroult, M. *Leibniz: Dynamique et Métaphysique.* Aubier-Montaigne, Paris, 1967.

[64] Gueroult, M. "Space, Point, and Void in Leibniz's Philosophy". In M. Hooker (ed.) *Leibniz: Critical and lnterpretative Essays,* University of Minnesota Press, Minneapolis, pp. 284-301, 1982.

[65] Haldane, E. and Ross, G. R. T. (eds. and trans.) *The Philosophical Works of Descartes.* Cambridge University Press, Cambridge, 1969.

[66] Hall, A.R. *Philoshophes at War.* Cambridge University Press, Cambridge, 1980.

[67] Hartz, G.A. "Leibniz's Phenomenalisms". *The Philosophical Review,* vol. 101, pp. 511-549, 1992.

[68] Hartz, G.A. and Cover, J. A. "Space and Time in the Leibnizian Metaphysic". *Noûs* 22, pp. 493-519, 1988

[69] Heller, M. and Staruszkiewicz, A. "A Physicist's View on the Polemics between Leibniz and Clarke". *Organon* vol. 11, pp. 205-213, 1975.

[70] Hooker, C.A. "Relational Doctrines of Space and Time". *British Journal for the Philosophy of Science,* vol. 22, pp. 97-130, 1971.

[71] Hooker, M. (ed.) *Leibniz. Critical and interpretative Essays.* University of Minnesota Press, Minneapolis, 1982.

[72] Hunter, G. "Monadic Relations". In K. Okruhlik and J. B. Brown (eds.) *The Natural Philosophy* of Leibniz, D. Reidel, Dordrecht, pp. 151-171, 1985.

[73] Ishiguro, H. *Leibniz's Philosophy of Logic* and *Language.* Cornell University Press, Ithaca, 1972.

[74] Ishiguro, H. "Leibniz's Theory of the Ideality of Relations". In H.G. Frarkfurt (ed.) *Leibniz: A Collection of Critical* Essays. University of Notre Dame Press, Notre Dame, pp. 191-2I3, 1972.

[75] Jagodinski , I.(ed.) *Leibnitiana elementa philosophiae arcanae de summa rerum.* Kasan, 1913.

[76] Jolley, N. *Leibniz and Locke: A study of the New Essays on Human Understanding.* Oxford University Press (Clarendon), Oxford, 1984

[77] Jolley, N. (ed.) *The Cambridge Companion to Leibniz.* Cambridge University Press, Cambridge 1995.

[78] Kant, I. "Concerning the Ultimate Foundation of the Differentiation of Regions in Space". In G.B. Kerferd and D.E. Walford, (eds. and trans.) *I. Kant: Selected Pre-Critical Writings and Correspondence with Beck* . Barnes and Noble, New York, pp. 36-43, 1968.

[79] Kant, I. *Critique of Pure Reason.* N. Kemp Smith (trans.) The McMillan Press LTD, London and Basingstoke, 1973.

[80] Kant, I. *Prolegomena to any Future Metaphysics.* J.W. Ellington (trans.) Hackett, 1977.

[81] Kant, I. *What Real Progress Has Metaphysics Made in Germany since the Time of Leibniz and Wolff?* T. Humphrey (trans.) Abaris Books, New York, 1983.

[82] Kleene, S.C. *Introduction to Metamathematics.* Wolters-Noorhoff, Groningen and North-Holland, London, 1971.

[83] Koyré, A. *Newtonian Studies.* University of Chicago Press, Chicago, 1965.

[84] Koyré, A. and Cohen, I.B. "Newton and the Leibniz Clarke Correspondence". *Archives internationales d' histoire des sciences,* vol. 15 pp. 63-126, 1962.

[85] Kreisel, G. and Krivine, J.L. *Elements of Mathematical Logic.* North Holland, Amsterdam, 1971.

[86] Kulstad, M. (ed.) *Rice University* Studies: *Essays on the Philosophy of Leibniz.* Rice University Press, Houston, 1977.

[87] Kulstad, M. "Some Difficulties in Leibniz's Definition of Perception". In M. Hooker (ed.) *Leibniz.- Critical and lnterpretative Essays,* University of Minnesota Press, Minneapolis, pp. 65-78, 1982.

[88] Kulstad, M. *Leibniz on Apperception, Consciousness and Reflection.* Philosophia Verlag, München, 1991.

[89] Kuratowski K. *Introduction to Set Theory and Topology.* Pergamon Press, Oxford and Polish Scientific Publishers, Warszawa 1972.

[90] Lacey, H. "The Philosophical Intelligibility of Absolute Space: A Study of Newtonian Argument". *British Journal for the Philosophy of Science,* vol. 21, pp. 317-342, 1971.

[91] Langley, A.G. (ed. and trans.) *New Essays Concerning Human Understanding, by Gottfried Wilhelm Leibniz: Together with an Appendix of his Shorter Pieces.* Second edition. Open Court Publishing Company, La Salle, 1916.

[92] Latta, R. (ed.and trans.) *Leibniz The Monadology and Other Philosophical Writings.* Oxford University Press, Oxford, 1898.

[93] Lear, J. "Aristotelian Infinity". *Proceedings of the Aristotelian Society* , vol. LXXX, pp. 187-210, 1979-1980.

[94] Loemker, L.E. (ed. and trans.) *Gottfried Wilhelm Leibniz: Philosophical Papers and Letters.* D. Reidel, Dordrecht, 1969.

[95] Lovejoy, A.O. *The Great Chain of Being.* Harvard University Press, Cambridge, 1964.

[96] Lucas. J.R. *Space Time and Causality.* Oxford University Press, Oxford, 1984.

[97] Lucas, P. G., and Grint, L. (trans.) *G.W. Leibniz: Discourse on Metaphysics.* Manchester University Press, Manchester, 1953.

[98] Luce, A.A. and Jessop, T.E. (eds.) *The Works of George Berkeley, Bishop of Cloyne.* In 9 volumes, Thomas Nelson and Sons Ltd., London, 1949.

[99] Mach, E. *The Science of Mechanics.* 9th ed. Open Court, London, 1942.

[100] Manders, K. "On the Space-Time Ontology of Physical Theories". *Philosophy of Science,* vol. 49, pp. 575-590, 1982.

[101] Martin, G. *Leibniz: Logic and Metaphysics.* K.J. Northcott and P.G. Lucas (trans.) Manchester University Press, Manchester, 1964.

[102] Mason, H.T. (ed. and trans.) *The Leibniz-Arnauld Correspondence.* Manchester University Press, Manchester, 1967.

[103] McGuire, J.E. "Labyrinthus Continui : Leibniz on Substance, Activity, and Matter". In P.K. Machamer and R. G. Turnbull (eds.) *Motion and Time, Space and Matter: lnterrelations in the History of Philosophy and Science.* Ohio State Uni versity Press, Columbus, pp. 290-326 , 1976.

[104] McGuire, J.E. "Existence, Actuality and Necessity: Newton on Space and Time". *Annals of Science,* vol. 35, pp. 463-508, 1978.

[105] McGuire, J.E. "Phenomenalism, Relations and Monadic Representation: Leibniz on Predicate Ievels". In J. Bogen and J.E. McGuire (eds.) *How Things Are.* D. Reidel, Dordrecht, pp. 205-233, 1985.

[106] McRae, R. *Leibniz: Perception, Apperception and Thought.* University of Toronto Press, Toronto, 1976.

[107] McRae, R. "The Theory of Knowledge". In N. Jolley (ed.) *The Cambridge Companion to Leibniz.* Cambridge University Press, Cambridge, pp. 176-198, 1995.

[108] Mendelson, E. *Introduction to Mathematical Logic.* Van Norstrand Reinhold Co., New York 1964.

[109] Mittelstrass, J. "Leibniz and Kant on Mathematical and Philosophical Knowledge". In K. Okruhlik and J. B. Brown (eds.) *The Natural Philosophy of Leibniz.* D. Reidel, Dordrecht, pp. 227-262, 1985.

[110] Mondadori, F. "Solipsistic Perception in a World of Monads". In M. Hooker (ed.) *Leibniz: Critical and Interpretative Essays.* University of Minnesota Press, Minneapolis, pp. 21-44, 1982.

[111] Montgomery, G.R (trans.) *Leibniz: Discourse on Metaphysics, Correspondence with Arnauld, Monadology.* Open Court, La Salle, 1980.

[112] Morris, M. (trans.) and Parkinson, G.H.R. (ed. and trans.) *Leibniz: Philosophical Writings.* Everyman's University Library, J.M. Dent and Sons, London, 1973.

[113] Munkres J.R. *Topology: A First Course.* Prentice Hall, Englewood Cliffs, New Jersey, 1975.

[114] Newton, I. *Mathematical Principles of Natural Philosophy.* A. Motte and F. Cajori (trans.) University of Californial Press, Berkeley, 1962.

[115] Okruhlik, K. and Brown, J.B. (eds.) *The University of Western Ontario Series in Philosophy of Science: The Natural Philosophy of Leibniz.* D. Reidel, Dordrecht, 1985.

[116] Parkinson, G.H.R *Logic and Reality in Leibniz's Metaphysics.* Oxford University Press, Oxford, 1965.

[117] Parkinson, G.H.R (ed. and trans.) *Leibniz Logical Papers. A Selection.* Oxford University Press, (Clarendon), Oxford, 1966.

[118] Parkinson, G.H.R "The Intellectualization of Appearances: Aspects of Leibniz's Theory of Sensation and Thought". In M. Hooker (ed.) *Leibniz: Critical and Interpretative Essays.* University of Minnesota Press, Minneapolis, pp. 3-20, 1982.

[119] Perl M.R. "Physics and Metaphysics in Newton Leibniz and Clarke". *Journal of the History of Ideas,* vol. 30, pp. 507-526, 1969.

[120] Perrett, W. and Jeffrey, G.B. (eds.) *The Principle of Relativity.* Dover, New York, 1952.

[121] Poincaré, H. *La Science et l' hypothèse,* Paris 1905. In English, *Science and Hypothesis*, Dover, 1952.

[122] Reihenbach, H. *Space and Time.* Dover, New York, 1957.

[123] Reihenbach, H. "The Theory of Motion according to Newton, Leibniz and Huyghens". Reprinted in M. Reihenbach (ed. and trans.) *Modern Philosophy of Science.* Routlege and Kegan Paul, London, 1959.

[124] Remnant, P. and Bennett, J. (eds. and trans.) *New Essays on Human Understanding.* Cambridge University Press, Cambridge, 1982.

[125] Rescher, N. *Leibniz : An Introduction to his Philosophy.* Basil Blackwell, Oxford, 1979.

[126] Rescher, N. *Leibniz Metaphysics of Nature: A Group of Essays.* Reidel, Dordrecht, 1982.

[127] Robinson, A. *Non-Standard Analysis.* North Holland Amsterdam, 1951.

[128] Russell, B. *A Critical Exposition of the Philosophy of Leibniz.* Eighth impression, G. Allen and Unwin Ltd., London, 1975.

[129] Russell, B. "Recent Work on the Philosophy of Leibniz". In H.G. Frankfurt (ed.) *Leibniz: A Collection of Critical Essays.* University of Notre Dame Press, Notre Dame, pp. 365-400, 1972.

[130] Russell, B. and Whitehead, A.N. *Principia Mathematica.* Cambridge University Press, London 1910.

[131] Russell, L.J. "The Correspondence Between Leibniz and De Volder". In R. Woolhouse (ed.) *Oxford Readings in Philosophy. Leibniz: Metaphysics and Philosophy of Science.* Oxford University Press, Oxford, pp. 104-118, 1981.

[132] Rutherford, D. "Metaphysics: the late period". In N. Jolley (ed.) *The Cambridge Companion to Leibniz.* Cambridge University Press, Cambridge, pp. 124-175, 1995.

[133] Schrecker, A.M. and Schrecker , P. (eds. and trans.) *Leibniz: The Monadology and Other Philosophical Essays.* Bobbs-Merrill, New York,1965.

[134] Schurle, A.W. *Topics in Topology.* North Holland, New York-Oxford, 1979.

[135] Sellars, W. "Time and the World Order". In H.Feigl, M. Scriven and G. Maxwell (eds.) *Minnesota Studies in the Philosophy of Science.* University of Minnesota Press, pp. 527-616, 1962.

[136] Sellars, W. "Meditations Leibniziennes". *American Philosophical Quarterly* vol. 2, pp.30-54,1965. Also in R Woolhouse (ed.) *Oxford Readings in Philosophy. Leibniz: Metaphysics* and *Philosophy of Science.* Oxford University Press,0xford, pp. 30-54, 1981.

[137] Sellars, W. "Berkeley and Descartes: Reflections on the Theory of Ideas". In P.K. Machamer and R.G. Turnbull (eds.) *Studies in Perception.* Ohio State University Press, Columbus, pp. 259-311, 1978.

[138] Sellars, W. "Foundation for a Metaphysics of Pure Process: The Carus Lectures of Wilfrid Sellars". *The Monist* vol. 64, pp. 1-90, 1981.

[139] Shapin, S. "Of Gods and Kings: Natural Philosophy and Politics in the Leibniz-Clarke Disputes". *Isis,* vol. 72, pp. 187-215, 1981.

[140] Sklar, L. *Space, Time and Space-Time.* University of California Press, Berkeley 1976.

[141] Sleigh, R.C., Jr. *Leibniz and Arnauld: A Commentary on their Correspandence.* Yale University Press, New Haven, 1990.

[142] Spivak, M. *Calculus.* Addison-Wesley, World Student Series Edition, W.A. Benjamin Inc., Reading, Massachusetts 1967.

[143] Stewart, L. "Samuel Clarke, Newtonianism, and the Factions of Post-Revolutionary England". *Journal of the History of Ideas,* vol. 42, pp. 53-72, 1981.

[144] Swoyer, C. "Leibnizian Expression". *Journal of the History of Philosophy,* vol. XXXIII, pp. 65-99, 1995.

[145] Tymienecka, T.A. *Leibniz's Cosmological Synthesis.* Von Gorcum, Assen, 1964.

[146] Van Fraasen, B.C. *An Introduction to the Philosophy of Time and Space.* Random House, New York, 1970.

[147] White, M. J. *The Continuous and the Discrete: Ancient Physical Theories from a Contemporary Perspective.* Oxford University Press (Clarendon), Oxford, 1992.

[148] Whitehead, A.N. *The Concept of Nature.* Cambridge University Press, Cambridge, 1971.

[149] Whitehead, A.N. *Process* and *Reality* Corrected by D.R Griffin and D.W. Sherburne. The Free Press, London, 1978.

[150] Whitehead, A.N. *An Enquiry Concerning the Principles of Natural Knowledge.* Dover Publications, New York, 1982.

[151] Whitt, L.A. "Absolute Space: Did Newton Take Leave of His (Classical) Empirical Senses?" *Canadian Journal of Philosophy* , vol. XII, pp. 709-724, 1982.

[152] Wiener, P.P. (ed.) *Leibniz: Selections.* Charles Scribner and Sons, New York, 1951.

[153] Wilson, C. *Leibniz's Metaphysics: A Historical and Comparative Study.* Princeton University Press, Princeton, 1989.

[154] Winterbourne, A.T. "On the Metaphysics of the Leibnizian Space and Time". *Studies in History and Philosophy of Science* , vol. 13, pp. 201-214, 1982.

[155] Woolhouse, R. (ed.) *Oxford Readings in Philosophy. Leibniz: Metaphysics* and *Philosophy of Science.* Oxford University Press, Oxford, 1981.

INDEX

A

"A New System of the Nature and the Communication of Substances as well as the Union between the Soul and the Body," 88
a priori, 50; 63; 73; 78; 97; 104
absolute container, 37; 91; 94; 96; 105; 106; 108; 134; 142
absolute motion, 106
absolutism, 107
action, 7; 8; 25; 47; 48; 82; 89
activity, 6; 7; 26; 28; 46; 47
actual infinite, 86; 136; 150; 159; 163; 166; 172
actual infinity, vii; 86; 87; 158
actuality, 87; 103; 162; 164
ad infinitum, 7; 9; 17; 22; 23; 24; 40; 42; 44; 65; 66; 67; 81; 87; 88; 94
algebraic numbers, 71
amnesia, 32
analysis, x; 13; 18; 19; 38; 43; 53; 54; 59; 83; 89; 92; 95; 121; 130; 131
animal, x; 27; 28; 29; 31; 32; 33; 34; 35; 36; 38; 39; 41; 67; 68; 73; 74; 75; 76; 77; 78; 82; 84; 90; 141
animal sensation, 29; 31
apperception, 23; 27; 28; 29; 30; 31; 36; 38; 141
appetite, 25; 27; 28; 36; 77
appetition, 7; 8; 29; 46; 47; 89; 127
architectonic principle, x; 50; 64
argument, x; 11; 51; 54; 55; 57; 58; 59; 80; 81; 106; 161; 162; 166
Aristotle, 66; 88; 91; 170; 174
arithmetic, 98
Arnauld, A., 11; 29; 146; 179; 181
assemblage, 7; 10; 21; 39; 43; 74
atom, 42; 58; 60; 61; 62; 63; 64; 66; 88; 89
auditory sense perception, 38
awareness, 23; 25; 29; 30; 32; 134
axiom, 53; 71; 81; 98; 155; 159

B

Bayle, P., 45; 146
beast, 68
be-causality, 3; 148
Berkeley, G., 11; 106; 179; 181
Bernoulli, J., 25; 45; 86
birth, 18; 32; 33; 83; 84
body, 3; 27; 28; 33; 34; 36; 45; 53; 59; 63; 65; 67; 68; 74; 75; 79; 82; 83; 97; 101; 103; 106; 108; 122; 125; 126; 127; 139; 163
boundary, 42; 43; 44; 46; 101; 103; 113; 164
Bourguet, L., 26; 76; 136; 138; 165

C

Cantor, G., 161; 175
cardinality, 61; 119; 125; 128; 143; 153; 162; 163; 172; 173
Cartesian tradition, ix
Cauchy sequence, 70; 71; 72; 73; 74; 77
Cauchy, A-L., x; 69; 70; 71; 72; 73; 74; 75; 76; 77; 174
causal process, 2
chain of being, 48; 66; 69; 77

change, vii; x; 7; 8; 25; 48; 50; 61; 62; 63; 64; 65; 67; 68; 85; 104; 105; 109; 113; 116; 131; 133; 134; 135; 136; 137; 138; 139; 140; 141; 142; 143; 144; 145; 146; 147; 148; 149; 150; 151; 152; 153; 154; 155; 156; 157; 158; 162; 167; 169; 170
circle, 66
Clarification of the Difficulties which Mr. Bayle has found in the New System of the Union of Soul and Body, 44; 146
Clarke, S., 9; 16; 57; 96; 99; 104; 105; 107; 108; 134; 135; 136; 138; 165; 174; 176; 177; 178; 180; 181
coexistence, 99; 100; 101; 104; 109; 116; 134; 166
Cohen, I. B., 107; 178
coherence, 5; 20; 81; 112; 140; 142
color, 5; 6; 111
communication, 12; 88; 126
complete concept, 6; 33; 44
complete notion, 58; 59
composition of the continuum, vii; x; 85; 88; 91; 92; 94; 95; 99; 102; 104; 117; 155; 158; 159; 162; 163; 166; 167; 168; 171
compossibility, 10
compound substance, 8
concept, ix; 6; 11; 12; 33; 44; 76; 96; 97; 98; 100; 135; 138; 146
conception, xi; 18; 73; 76; 77; 107
confused perceptions, x; 22; 26; 31; 32; 33; 35; 38; 68; 124
confusedness, x; 15; 19; 26; 27; 29; 30; 31; 32; 33; 37; 44; 46; 47; 48; 66; 69; 75; 114; 117; 122; 123; 127; 130; 170
conscious perceiving, 28
consciousness, 28; 30; 100
consecutiveness, 34; 141
continuity, vii; ix; x; 22; 37; 39; 40; 42; 48; 50; 51; 52; 53; 54; 55; 56; 57; 58; 59; 60; 62; 63; 64; 65; 66; 67; 68; 69; 70; 72; 73; 74; 75; 76; 77; 78; 79; 102; 115; 116; 117; 118; 121; 124; 127; 131; 137; 149; 150; 153; 158; 169; 170
continuity of actual existents or forms, 65; 66
continuity of cases, 65; 66
continuity of existence *ad infinitum*, 65; 66; 67
continuity of space and time, 65; 66
continuous, ix; x; xi; 7; 22; 27; 29; 30; 31; 32; 33; 36; 39; 42; 48; 50; 51; 52; 53; 54; 55; 56; 57; 59; 62; 64; 65; 66; 67; 68; 69; 72; 73; 74; 75; 76; 78; 85; 87; 92; 102; 103; 111; 115; 116; 118; 119; 120; 125; 131; 132; 135; 136; 137; 138; 139; 147; 148; 149; 150; 151; 153; 157; 162; 164; 165; 167; 168; 169; 170; 171; 173
continuous function, x; 51; 52; 53; 66; 74; 75
continuous grading of monads, 32
continuous spectrum, 27; 29; 30; 31; 36; 37; 48; 66; 73; 76
continuum, vii; x; 59; 70; 73; 75; 78; 79; 83; 85; 87; 88; 90; 91; 92; 94; 95; 97; 99; 102; 103; 104; 109; 115; 116; 117; 118; 119; 120; 128; 132; 135; 136; 137; 139; 149; 153; 155; 158; 159; 162; 163; 164; 165; 166; 167; 168; 169; 171; 172; 173
contradiction, 56; 161; 164
convergence, 70; 71
convergent sequence, 71
correspondence, ix; 2; 3; 4; 6; 8; 10; 20; 22; 30; 43; 45; 57; 61; 75; 97; 107; 110; 111; 119; 120; 121; 122; 128; 161

Correspondence with Arnauld, 146; 179
corruption, 67
creation, 10
creator, 9
criterion, 5; 20; 54; 111; 112; 167

D

data, 50; 51; 52; 66; 74
David, 12; 13; 14; 15; 16; 18; 19
death, 28; 32; 33; 67; 83; 84
Dedekind cuts, 70
Dedekind, R., 70
definition, 26; 171
degrees of distinctness or confusedness, 29
degrees of perfection, 33; 47
density, vii; x; xi; 22; 39; 40; 42; 48; 55; 67; 69; 70; 71; 72; 73; 76; 77; 103; 110; 111; 117; 120; 121; 131; 132; 133; 135; 137; 144; 150; 151; 152; 153; 155; 156; 157; 167; 169; 170; 171; 173
Des Bosses, B., 12; 25; 82; 85; 97; 99; 101; 116; 134; 160; 169
Des Maizeaux, P., 27
Descartes, R., 1; 3; 4; 5; 6; 89; 174; 175; 177; 181
direct representation, 14; 21; 22; 40; 148; 156
directionality, 135; 138; 149; 156; 165
discontinuity, vii; x; 63; 68
discontinuous, 51; 52; 54; 56; 57; 68; 118; 119; 146; 147; 148; 150
Discourse on Metaphysics, 11; 48; 58; 149; 159; 179
discrete, 8; 56; 60; 61; 72; 83; 85; 87; 115; 116; 117; 118; 119; 131; 146; 147; 148; 150; 151; 153; 163; 167; 170
Discrete and Continuous Model, xi; 151
Discrete and Discontinuous Model, xi; 146
discreteness, 106; 115; 116; 131; 171
Dissertatio de arte combinatoria, 79; 92
distance, 61; 68; 69; 103; 104; 105; 113; 114; 125; 126; 127; 144; 152; 156; 173
distinct perceptions, x; 23; 26; 30; 31; 32; 34; 35; 38; 68; 126; 127
distinctness, x; 15; 26; 27; 29; 30; 31; 32; 33; 34; 36; 38; 44; 46; 47; 48; 66; 69; 75; 110; 114; 117; 122; 123; 125; 127; 130; 140; 156; 170
doctrine, x; 6; 8; 9; 30; 49; 78; 81; 87; 91; 92; 94; 104; 128; 139; 141; 159; 162; 167
domain, 1; 19; 51; 52; 53; 86
dominant monad, 15; 27; 31; 33; 34; 35; 36; 43; 66; 67; 73; 75; 127
dyadic relation, 15

E

Earman, J., 14; 107; 175
Einstein, A., 106; 107; 175; 176
ellipse, 64
ens rationis, 134
entelechy, 32; 33; 46; 126; 127
entity, x; 4; 5; 6; 41; 60; 69; 73; 79; 83; 84; 87; 90; 91; 94; 95; 96; 97; 98; 100; 101; 102; 103; 104; 108; 113; 115; 117; 121; 128; 129; 131; 133; 134; 135; 136; 142; 144; 145; 156; 162; 164; 167; 168; 170; 171
epistemology, 26
equality, 63; 64; 66; 73; 128; 153; 162
essence, 1; 3; 11; 12; 118
Euclidean geometry, 2; 73; 95; 98; 101; 106; 139
Euclidean space, 2
Euclidean world, 2
Euler, L., 106
explanation, 3; 5; 6; 40; 52; 53; 75; 112

expression, 2; 14; 27; 29; 30; 32; 48; 52; 65; 73; 90; 101; 123; 124; 128; 129; 130; 156
extended, vii; x; 5; 22; 37; 38; 39; 40; 41; 46; 48; 50; 63; 64; 65; 68; 69; 70; 72; 79; 81; 82; 83; 84; 85; 86; 87; 88; 90; 91; 92; 94; 95; 97; 98; 101; 102; 103; 106; 109; 111; 113; 115; 118; 119; 120; 121; 122; 128; 131; 132; 136; 137; 138; 140; 142; 145; 150; 159; 162; 167; 168; 170; 171
extension, 5; 6; 55; 72; 78; 85; 92; 95; 96; 97; 100; 102; 111; 122; 134; 138; 160; 162; 164; 166; 167; 169; 172
external reality, 43
external world, 11; 17; 20

F

feeling, 24; 32; 80; 127; 142
finite, 96; 161
finite substance, 48
finite world, 55
finitude, x; 55; 56
first-order relation, 14
form, x; 2; 4; 5; 7; 10; 13; 15; 16; 19; 28; 29; 30; 33; 36; 37; 38; 40; 45; 48; 50; 51; 52; 53; 55; 59; 62; 65; 66; 69; 73; 74; 77; 88; 89; 91; 100; 104; 108; 109; 114; 115; 119; 120; 121; 124; 128; 129; 130; 132; 133; 137; 139; 142; 146; 153; 158; 162; 170; 171; 173
formal reality, 3; 4; 6; 8; 9; 10; 14; 20; 25; 37; 122; 157
formal realization, 2; 3; 11
formal/objective dichotomy, 19; 20
formalization, vii; xi; 70; 121; 153; 155; 157
formula, 133
fossils, 36; 74; 75
Foucher, S, 86; 100; 159; 176
foundation, 39; 57; 58; 138; 144; 148
framework, ix; x; 4; 6; 17; 20; 21; 26; 61; 62; 68; 70; 96; 101; 102; 103; 106; 108; 110; 113; 120; 132; 140; 156; 164
function, x; 51; 52; 53; 66; 74; 75; 129; 143; 172; 173
future, xi; 2; 7; 25; 34; 44; 45; 73; 112; 121; 137; 138; 140; 141; 143; 146; 148; 149; 156; 157; 164

G

generation, 66; 67
genuine continuity, xi; 150
genuine continuum, 149; 166; 167; 168; 171
genus, 29
geometry, 2; 73; 90; 95; 98; 99; 101; 106; 139
God, 1; 3; 4; 7; 9; 10; 11; 12; 17; 19; 31; 33; 44; 46; 47; 48; 53; 54; 55; 57; 58; 59; 74; 76; 96; 97; 110; 114; 115; 118; 138; 141; 142; 145; 146; 150; 157; 158; 159; 160; 163; 170; 176
God's free will, 10; 11; 55
God's understanding, 10

H

Hall, A. R., 107; 177; 179
hiatus, 60
homogeneity, 135; 163; 165; 171
Huygens, C., 106; 107

I

idea, xi; 2; 4; 5; 9; 10; 11; 21; 23; 42; 48; 51; 52; 60; 62; 70; 73; 74; 79; 80; 82; 83; 84; 92; 94; 97; 98; 99; 100; 103; 104; 106; 108; 131; 132; 134; 143; 147;

148; 159; 160; 161; 162; 164; 166; 168; 170
ideal, 6; 103; 135; 163; 165
ideal continuum, x; 135; 136; 162; 164; 165; 166
ideal entity, x; 6; 91; 95; 96; 97; 98; 100; 101; 102; 103; 104; 108; 134; 135; 136; 142; 145; 168; 171
ideal level, 49; 85; 171
identity of indiscernibles, x; 58; 59
immutability, 108
imperfection, 31; 48; 53; 55; 56; 74; 96
impression, 32; 34; 35; 56; 68; 69; 109; 116; 142; 146; 149; 170; 180
indirect representation, ix; xi; 7; 9; 14; 15; 20; 21; 22; 23; 24; 29; 39; 40; 41; 43; 110; 120; 122; 124; 125; 140; 156; 173
indiscernible, x; 58; 59; 111
individual substance, 7; 10; 12; 146; 147; 159
indivisible substance, 8; 13; 80
inequality, 63; 64; 66; 73
infinite, 8; 14; 19; 22; 23; 24; 31; 34; 38; 39; 40; 42; 45; 46; 50; 54; 55; 56; 57; 60; 64; 66; 69; 72; 79; 81; 82; 85; 86; 87; 88; 90; 92; 94; 95; 96; 103; 109; 113; 114; 116; 117; 119; 120; 121; 124; 126; 127; 128; 131; 135; 136; 149; 150; 151; 152; 153; 158; 159; 160; 161; 162; 164; 166; 167; 168; 172
infinite divisibility, 38; 42; 50; 66; 79; 81; 82; 88; 90; 92; 95; 117; 128; 131; 135; 136; 150; 166; 168
infinite number, 79; 85; 86; 87; 116; 160; 161; 163; 166
infinite regress, 23; 24; 95
infinite world, 54; 55; 56; 57; 60
infinitesimal, 66; 84; 90; 107
infinitesimal calculus, 107
infinitude, 22; 24; 25; 31; 37; 38; 39; 40; 56; 73; 85; 96; 114; 124; 137; 151; 152; 159; 160; 162; 166; 167; 169; 171; 172
infinity, vii; 24; 26; 27; 28; 29; 39; 43; 75; 86; 87; 88; 114; 117; 126; 152; 158; 160
influxus physicus, 9
innate idea, 97; 98; 108; 134
insensible perceptions, 30; 38; 42; 43; 44
integers, 60; 61; 71; 72
intellect, 53
intellectual knowledge, 29; 31
intellectual satisfaction, 54
interpretation, ix; xi; 13; 17; 18; 60; 120; 121; 126; 129; 130; 131; 144; 145; 146; 150; 153
intersubjectivity, 112; 141; 142
inter-monadic change, 140; 142
intra-monadic change, 140
isomorphism, 68; 95; 110; 122; 144

J

Johann Friedrich, Duke of Braunchweig-Lüneburg, 82
Journal des savans, 88; 100; 159
joy, 48

K

Kant, I., 106; 174; 176; 178; 179
kernel, 82; 83
knowledge, 4; 26; 29; 30; 31; 46; 97; 99; 103; 114; 117; 126; 158
Koyré, A., 107; 178

L

labyrinth of the continuum, 78; 85; 87; 90; 94; 116; 163; 164; 165

labyrintus continui, xi
Landgrave of Hessen-Rheinfels, E., 48
law, 50; 53; 68; 115
leap, x; 50; 51; 53; 54; 62; 64; 66; 67; 68; 169
Leibniz-Clarke Correspondence, The, 9; 16; 57; 96; 99; 104; 107; 108; 134; 136; 138; 174
Leibniz's metaphysics, ix; 90; 94; 148; 150; 153
Leibniz's system, ix; x; 63; 110; 128; 131
Leibniz's writings, x; 51; 57; 65; 74; 102; 116
Leibnizian doctrine, x; 9; 91; 104; 115; 128; 139; 167
Leibnizian metaphysics, 8; 48; 53; 64; 145; 170
lethargy, 30; 32
Letter of Mr. Leibniz on a General Principle Useful in Explaining the Laws of Nature Through a Consideration of the Divine Wisdom; to Serve as a Reply to the Response of the Rev. Father Malebranche, 63
level, ix; x; 6; 7; 37; 49; 53; 59; 64; 72; 75; 85; 88; 91; 92; 94; 97; 100; 102; 106; 109; 111; 112; 115; 117; 118; 119; 120; 131; 132; 133; 140; 141; 145; 148; 150; 151; 154; 156; 157; 163; 165; 166; 167; 168; 169; 170; 171
limit, 31; 51; 66; 67; 71; 73; 74; 77; 132; 164
line, 21; 30; 31; 51; 52; 65; 69; 70; 71; 72; 73; 85; 87; 91; 103; 109; 119; 123; 125; 132; 135; 139; 143; 144; 148; 151; 160; 162; 165; 166; 170; 171; 172; 173
line segment, 30; 31; 51; 65; 70; 73; 85; 143; 144; 151; 162; 173
linearity, 110; 139; 157
linearly continuous spectrum of beings, 36
Lovejoy, A. O., 60; 178

M

Mach, E., 106; 179
magnitude, 42; 83; 84; 85; 90; 96
Malebranche, N., 50; 63; 160
mass, 24; 35; 38; 39; 63; 89; 116; 149
materiality, 85
mathematical point, 82; 83; 87; 88; 89; 90; 95; 101; 109; 162
mathematics, 51; 57; 86; 100; 105; 108; 132
matter, x; 2; 3; 13; 21; 24; 25; 31; 33; 39; 40; 52; 73; 81; 82; 83; 84; 85; 86; 87; 88; 89; 98; 99; 103; 111; 114; 115; 116; 123; 124; 125; 126; 131; 135; 137; 146; 150; 162; 166; 167
maxim, 62; 66; 161; 164; 169
Maxwell, J., 106; 180
McGuire, J. E., xi; 44; 179
McRae, R., 38; 179
McTaggard, J. M. E., 58
measurement, 104; 105; 140; 143; 152; 162
mechanics, 63; 106
mechanism, 2; 3; 33; 34; 36
memory, 2; 30; 31; 34; 45; 140; 141
mental state, 3; 27
mental thing, 12; 17
metaphysical, 19; 22; 59; 60; 65; 73; 112; 166
Metaphysical Foundations of Mathematics, 150; 173
metaphysical level, 59; 88; 132; 133; 141; 145; 150; 151
metaphysical point, 83; 87; 89; 90; 162; 172
metaphysical simultaneity, xi; 141; 143

metaphysical vacuum, 83; 84
metaphysician, 17; 55; 56; 59; 81; 170
metaphysics, ix; x; 4; 5; 6; 8; 9; 21; 26; 41; 44; 48; 53; 64; 80; 90; 94; 95; 112; 117; 145; 148; 150; 153; 158; 159; 161; 162; 163; 168; 169; 170; 171
metempsychosis, 67
microscope, 27; 38; 42; 43; 90
mind, 1; 2; 3; 4; 9; 10; 11; 13; 24; 28; 30; 33; 38; 42; 43; 44; 76; 79; 82; 83; 84; 86; 95; 97; 98; 100; 104; 109; 116; 128; 140; 150; 158; 159
minimum perceptibile, 42; 43; 90; 113
minute perceptions, 29; 42
model, vii; x; 20; 21; 22; 23; 40; 43; 58; 73; 79; 95; 102; 109; 111; 118; 119; 120; 123; 124; 125; 133; 134; 140; 141; 142; 144; 145; 146; 148; 150; 151; 153; 155; 156; 157; 170
monad, ix; 7; 8; 11; 12; 13; 14; 15; 17; 18; 20; 21; 22; 23; 25; 26; 27; 28; 29; 30; 31; 32; 33; 34; 35; 36; 37; 39; 40; 41; 43; 44; 46; 47; 48; 56; 58; 59; 60; 66; 67; 68; 72; 73; 75; 78; 81; 86; 87; 89; 91; 97; 105; 108; 109; 110; 111; 113; 114; 115; 117; 118; 119; 120; 121; 122; 123; 124; 125; 126; 127; 128; 129; 130; 131; 132; 133; 136; 137; 138; 140; 141; 142; 143; 145; 147; 148; 150; 151; 152; 153; 154; 155; 156; 157; 159; 161; 163; 165; 166; 167; 169; 170; 171; 172; 173
monadic change, vii; xi; 105; 109; 133; 134; 137; 140; 141; 142; 144; 145; 146; 147; 148; 150; 151; 152; 156; 157; 169; 170
monadic metaphysics, x; 44; 161; 168; 171
monadic reality, x; 115; 117; 141; 143; 170
monadic state, 8; 20; 133; 144; 145; 151; 152; 153; 169
Monadology, The, 8; 25; 26; 28; 34; 46; 48; 56; 125; 127; 140; 146; 150; 152; 157; 169
mortality of the soul, 28
motion, 50; 53; 63; 64; 66; 73; 97; 106; 107; 108; 126; 127; 134; 152
Munkres, J. R., 52; 179

N

natural perception, 29; 30; 31
natural philosophy, 57
nature, x; 2; 3; 4; 6; 11; 14; 15; 18; 23; 31; 32; 33; 45; 46; 49; 50; 51; 52; 53; 55; 56; 61; 62; 66; 68; 69; 74; 77; 79; 86; 87; 88; 94; 98; 99; 100; 106; 107; 111; 122; 126; 136; 137; 138; 139; 140; 142; 144; 145; 146; 147; 150; 151; 158; 159; 162; 165; 169
nature never makes leaps, x; 50; 51; 62; 66; 169
nature respects limits, 66
necessity, 50; 138; 158
New Essays on Human Understanding, 9; 24; 28; 32; 36; 47; 50; 68; 76; 86; 97; 98; 139; 160; 177; 178; 180
New System of the Nature and of the Union of the Soul and Body, 100
Newton, I., 106; 107; 178; 179; 180; 181
nonconventionalism, 108
''Note on Foucher's objection,'' 100
notion, ix; x; xi; 9; 16; 21; 22; 38; 39; 48; 51; 52; 61; 70; 79; 83; 86; 95; 102; 103; 104; 106; 108; 122; 136; 138; 146; 151; 152; 162; 163; 166; 171
Nouvelles de la république des lettres, 50

O

object, 3; 4; 5; 9; 10; 11; 16; 26; 35; 37; 40; 46; 60; 64; 65; 68; 72; 84; 86; 88;

90; 91; 94; 95; 96; 102; 104; 106; 108; 109; 113; 125; 126; 127; 131; 132; 136; 167
objective reality, ix; 3; 4; 5; 6; 7; 8; 9; 18; 19; 25
objectivity, 112; 141
occasionalism, 9
On Freedom, 158
On The Method of Distinguishing Real from Imaginary Phenomena, 20; 111
ontological causality, 3
order, 9; 14; 15; 20; 21; 24; 25; 26; 30; 34; 39; 48; 51; 52; 53; 54; 56; 57; 60; 62; 63; 64; 67; 69; 71; 72; 74; 77; 81; 83; 84; 86; 89; 90; 93; 99; 100; 101; 103; 104; 105; 106; 109; 111; 116; 121; 122; 128; 131; 132; 134; 135; 139; 142; 143; 150; 151; 153; 160; 161; 163; 164; 165; 166; 167; 173
organism, 33; 67; 83; 84

P

pain, 34; 47; 48
parabola, 64; 66
Paris Notes, 82; 83; 85
passion, 47; 48
passivity, 46; 47
past, xi; 7; 34; 44; 45; 60; 112; 137; 138; 140; 141; 143; 148; 156
perceiver, 29; 31; 32; 34; 37; 41; 44; 46
perceiving, 3; 27; 28; 29; 31; 34; 36; 38; 40; 68; 124; 130
perception, vii; 2; 4; 8; 25; 26; 27; 28; 29; 30; 31; 32; 34; 36; 38; 40; 42; 45; 46; 47; 77; 89; 109; 113; 132; 141
perceptual state, 7; 27; 31; 33; 37; 47; 151
perfect point, 83; 84
perfection, 47; 48; 86
perspective, 7; 8; 21; 26; 29; 31; 44; 60; 89; 90; 109; 110; 114; 115; 118; 119; 120; 122; 123; 126; 127; 130; 133; 137; 138; 152; 157
perspective line, 21; 109; 123
perspective representation, 21; 109; 110; 118; 119; 120; 127; 130; 133; 152
perspective viewing, 8; 111; 118; 127; 138
petite perceptions, ix; 31; 34; 35; 37; 38; 4
phenomenal, vii; ix; x; 6; 12; 13; 14; 16; 18; 19; 20; 21; 22; 23; 25; 35; 37; 39; 40; 41; 42; 43; 44; 46; 49; 53; 59; 60; 61; 62; 64; 68; 72; 75; 81; 88; 91; 94; 97; 101; 103; 105; 108; 109; 110; 111; 115; 117; 118; 119; 120; 121; 122; 123; 124; 125; 126; 128; 129; 131; 132; 133; 134; 135; 136; 137; 138; 139; 140; 141; 142; 143; 144; 145; 146; 147; 148; 149; 151; 152; 153; 154; 155; 156; 157; 163; 166; 167; 168; 171; 172
phenomenal change, vii; xi; 133; 134; 135; 136; 137; 138; 139; 140; 141; 142; 143; 144; 145; 147; 149; 151; 152; 153; 155; 156; 157; 167
phenomenal level, ix; 37; 59; 72; 75; 88; 92; 105; 120; 140; 141; 151; 157
phenomenal realm, 18; 43; 103
phenomenal relation, ix; 14; 91; 108
phenomenal simultaneity, xi; 141
phenomenon, 5; 6; 8; 12; 13; 15; 20; 33; 39; 40; 41; 43; 44; 56; 60; 65; 68; 72; 73; 81; 88; 90; 93; 95; 96; 97; 99; 100; 101; 103; 106; 108; 109; 110; 111; 112; 113; 114; 115; 117; 118; 120; 121; 122; 132; 134; 135; 136; 137; 138; 142; 148; 156; 157; 163; 165; 166; 167; 170
phenomenon *bene fundatum*, 13; 98; 118; 120; 167

philosophy, 1; 3; 57; 74; 78; 82; 94; 115; 123; 159
physical plenum, 83; 84; 114; 167
physical points, 82; 83; 84; 85; 90
physics, 2
pineal gland, 2
place, 1; 2; 12; 18; 26; 43; 50; 53; 59; 65; 67; 74; 82; 86; 88; 89; 101; 104; 117; 135; 140; 148; 151; 163; 165
plant, 36; 75; 77
Plato, 97
Platonic Forms, 1
pleasure, 47; 68
plenum, 40; 58; 60; 62; 68; 69; 83; 84; 96; 114; 118; 119; 126; 127; 132; 167; 171
Poincaré, J. H., 106; 107; 180
point, 3; 6; 8; 9; 10; 12; 14; 19; 20; 21; 22; 26; 28; 30; 31; 37; 38; 41; 43; 44; 45; 47; 51; 52; 60; 61; 65; 66; 67; 68; 69; 70; 71; 72; 73; 74; 76; 77; 78; 79; 81; 82; 83; 84; 85; 87; 88; 89; 90; 92; 95; 99; 101; 102; 103; 106; 107; 108; 109; 110; 115; 117; 118; 119; 124; 127; 128; 131; 132; 135; 136; 138; 141; 142; 144; 145; 146; 148; 156; 157; 162; 164; 165; 166; 168; 169; 170; 172; 173
point of view, 6; 20; 26; 60; 66; 69; 81; 87; 89; 108; 109; 110; 115; 127; 135; 141; 142; 146; 157; 165; 170
point-of-viewish representations, 9; 45
polyadic relations, 15
polynomial equation, 72
position, x; 5; 12; 14; 15; 30; 35; 38; 41; 44; 57; 59; 60; 61; 65; 83; 87; 91; 95; 101; 103; 104; 106; 108; 111; 119; 120; 138; 142; 152; 153; 157; 165; 172
positional predicates, 105; 110; 172
positional structure, 172
possibilia, 10; 59
postulate, 4; 56; 59; 61; 77; 140
potentiality, 33; 65; 72; 94; 99; 102; 164
predicate, 12; 13; 15; 21; 44; 58; 59; 71; 92; 105; 108; 110; 128; 144; 152; 154; 172
predicative inherence, 16
pre-established harmonious agreement, 9
pre-established harmonious coordination, 17; 141
pre-established harmony, 11; 45; 141; 151; 157
present, x; xi; 3; 5; 9; 11; 19; 20; 23; 24; 25; 26; 33; 34; 45; 46; 63; 65; 67; 69; 82; 83; 112; 125; 126; 134; 138; 140; 141; 146; 147; 153; 154; 156; 157; 158; 172
prima facie, 13; 14; 18; 20; 21; 122
primary qualities, 5
primitive entelechy, 32; 33
principle of continuity, x; 22; 37; 42; 48; 50; 51; 52; 53; 54; 55; 56; 57; 58; 59; 60; 62; 63; 64; 65; 66; 67; 68; 69; 74; 76; 78; 117; 124; 127; 137; 150; 153; 169
principle of contradiction, 56; 57
principle of the best, x; 53; 55; 56; 58; 169
principle of general order, 53; 54; 64
principle of plenitude, x; 22; 58; 59; 60; 62; 127
principle of sufficient reason, x; 56; 57; 58; 169
principle of the identity of indiscernibles, x; 58; 59
Principles of Nature and of Grace Based on Reason, The, 8; 25; 34
problem, vii; ix; x; 3; 11; 13; 21; 22; 23; 24; 40; 44; 49; 55; 59; 78; 84; 85; 86; 88; 90; 91; 94; 95; 102; 117; 131; 134;

140; 143; 147; 150; 155; 158; 159; 162; 163; 165; 167; 168; 170; 171; 172
problem of the composition of the continuum, x; 88; 91; 94; 102; 117; 155; 158; 159; 162; 163; 167; 168; 171
process, 2; 43; 66; 80; 106; 108; 132; 143; 164
property, vii; x; 3; 5; 6; 12; 14; 18; 22; 39; 41; 51; 55; 66; 69; 70; 71; 73; 74; 75; 76; 77; 85; 95; 107; 117; 121; 135; 137; 139; 145; 149; 156; 162; 167; 168; 170; 171
proposition, 13; 56; 57; 98; 99; 146
psychophysics, 42
puzzle, 87; 92; 113; 132; 162

Q

quality, ix; 3; 5; 8; 25; 33; 111
quantity, 50; 87; 103; 105; 160; 163; 168

R

range, 51; 52; 133; 153; 157
rational numbers, 71
rational soul, 30; 33; 34; 36
real, ix; xi; 1; 5; 6; 7; 8; 11; 13; 14; 17; 18; 20; 21; 25; 37; 39; 41; 43; 49; 51; 53; 56; 59; 61; 62; 64; 65; 68; 69; 70; 71; 72; 73; 75; 77; 78; 79; 80; 83; 84; 85; 86; 87; 88; 89; 90; 91; 93; 94; 95; 96; 97; 98; 108; 109; 110; 111; 112; 113; 114; 115; 116; 117; 118; 119; 120; 121; 122; 123; 126; 128; 129; 130; 131; 132; 133; 135; 140; 141; 142; 144; 146; 148; 150; 151; 152; 153; 154; 155; 156; 157; 160; 161; 162; 164; 165; 166; 167; 168; 169; 170; 171; 172; 173
real infinite, 86; 162
real level, 92; 117; 120; 163
real line, 51; 69; 70; 71; 72; 73; 119; 132; 172
real numbers, 51; 70; 71; 77; 153
reality, ix; x; 3; 4; 5; 6; 7; 8; 9; 10; 11; 12; 14; 15; 18; 19; 20; 25; 31; 37; 39; 43; 65; 68; 80; 81; 88; 90; 91; 92; 105; 110; 111; 112; 113; 114; 115; 117; 118; 119; 121; 122; 124; 125; 141; 142; 157; 167; 170
realization, 1; 2; 3; 5; 11; 44; 132
reals, 72; 86
reason, x; 23; 33; 34; 40; 48; 50; 53; 54; 55; 56; 57; 58; 64; 89; 96; 98; 100; 111; 112; 124; 125; 131; 141; 142; 143; 150; 158; 161; 162; 163; 169
reductio ad absurdum, 57
reduction, ix
reflection, 23; 24; 29; 32; 80
Reflections on the Doctrine of a Single Universal Spirit, 42
relation, vii; ix; xi; 8; 12; 13; 14; 15; 16; 17; 18; 19; 20; 21; 22; 23; 38; 46; 50; 51; 53; 55; 58; 69; 91; 92; 100; 101; 103; 104; 105; 106; 108; 111; 113; 114; 117; 119; 120; 121; 122; 123; 124; 126; 128; 129; 131; 134; 143; 146; 154; 162; 163; 172; 173
relation of spatially between, 111
relational bridge, 17; 91
relational statement, 12; 13; 15; 16; 18; 19; 23; 121
relational theory of space, 91
relationism, 107; 108
Remond, N., 26; 67; 99; 134; 166
Reply to the Thoughts of the System of Preestablished Harmony Contained in the Second Edition of Mr. Bayle's Critical Dictionary, Article Rorarius, 45
representable, 1; 2; 3; 4; 7; 9; 10; 11

representation, vii; ix; x; xi; 2; 3; 4; 7; 8; 9; 11; 12; 13; 14; 15; 17; 18; 19; 20; 21; 22; 23; 24; 25; 26; 27; 29; 30; 31; 32; 34; 37; 39; 40; 41; 43; 45; 46; 48; 65; 68; 69; 71; 72; 73; 75; 83; 90; 91; 97; 105; 109; 110; 113; 114; 115; 118; 119; 120; 122; 124; 126; 127; 128; 130; 131; 132; 133; 134; 135; 136; 137; 138; 140; 141; 142; 143; 144; 145; 148; 149; 151; 152; 154; 156; 157; 160; 167; 168; 170; 171; 172; 173
representational act, 19; 27; 30
representational concepts, ix
representational density, xi
representational fact, ix; 16; 18; 121; 122; 123; 128; 129; 130; 132; 133; 173
representational framework, ix
representational metaphysics, ix; xi; 4; 5; 6; 21; 41; 44; 48
representational reality, x; 111
representational state, ix; 6; 7; 8; 9; 13; 14; 15; 17; 18; 20; 25; 27; 31; 33; 35; 36; 37; 47; 89; 110; 118; 120; 121; 122; 125; 128; 129; 130; 131; 132; 133; 134; 137; 140; 141; 143; 147; 148; 151; 152; 153; 172
representational structure, x; 15; 18; 19; 33; 40; 42; 43; 44; 59; 66; 67; 69; 75; 89; 105; 109; 113; 114; 117; 118; 119; 123; 127; 131; 152; 156; 170; 171; 172; 173
representational theory of truth, 3
representationalism, 1; 2; 5; 7; 49
representationally between, 123; 128; 130; 132; 133; 173
representing, ix; 1; 6; 7; 8; 11; 12; 13; 14; 15; 17; 18; 19; 20; 21; 22; 23; 24; 25; 29; 37; 39; 40; 43; 44; 45; 46; 59; 60; 65; 68; 69; 81; 88; 91; 105; 108; 109; 110; 113; 114; 115; 117; 118; 119; 121; 122; 123; 124; 125; 126; 129; 130; 131; 133; 137; 140; 141; 143; 144; 150; 151; 154; 156; 157; 165; 167
representor, 2; 7; 9; 17; 18; 19; 22; 24; 30; 41; 109; 115; 125; 126; 128; 132; 133; 153; 155; 156
Rescher, N., xi; 7; 10; 26; 162; 165; 180
rest, ix; 3; 8; 14; 24; 31; 64; 66; 73; 90; 122; 127; 152
Robinson, A., 84; 180
Russell, B., 42; 46; 58; 59; 65; 66; 78; 87; 89; 115; 117; 160; 180

S

secondary qualities, ix; 5
segment, 70
self-awareness, 25; 29; 30
self-reflection, 23; 24; 25; 29
sequential or Cauchy completeness, 69; 70; 71; 72; 73; 75; 77; 78
Sellars, W., ix; xi; 3; 13; 18; 121; 180; 181
sensation, 5; 27; 28; 29; 30; 31; 34; 35; 36; 38; 43
sense, xi; 1; 3; 4; 5; 6; 7; 8; 10; 13; 14; 17; 19; 21; 22; 23; 25; 27; 28; 33; 35; 37; 38; 40; 42; 43; 44; 46; 47; 49; 51; 52; 54; 56; 58; 59; 60; 64; 65; 67; 68; 69; 70; 72; 73; 74; 77; 78; 84; 91; 92; 96; 97; 98; 100; 101; 102; 103; 106; 108; 109; 111; 112; 113; 115; 116; 117; 118; 120; 121; 124; 127; 128; 131; 132; 134; 135; 142; 149; 150; 151; 153; 157; 159; 160; 163; 164; 167; 168; 169; 170; 171; 172; 173
sensing, 34; 35; 42; 125
sensory perception, 42
sentient life, 33
sentiment, 34

separation, 52; 65; 74; 108
sequence, 22; 40; 45; 63; 70; 71; 72; 74; 77; 112; 124
set, 8; 22; 51; 52; 60; 62; 69; 70; 71; 72; 74; 77; 100; 104; 107; 108; 110; 111; 112; 119; 121; 123; 125; 127; 129; 130; 138; 143; 145; 159; 162; 163; 165; 172; 173
shape, 5; 6; 34; 64; 113
simple monad, 27; 32; 33
simple substance, 6; 7; 8; 25; 27; 28; 32; 44; 81; 87; 88; 89; 127; 146; 157; 160; 162; 163
simultaneity, xi; 108; 109; 140; 141; 142; 143; 145; 151; 157
sleep, 32
Solomon, 12; 13; 14; 15; 16; 18; 19
solution, vii; xi; 3; 21; 24; 40; 85; 87; 90; 93; 142; 144; 155; 158; 159; 162; 163; 167; 168; 171
Sophia, Electress of Hannover, 149
sorites problem, 162; 168; 172
soul, 3; 15; 28; 30; 33; 34; 35; 36; 38; 45; 46; 67; 68; 82; 83; 85; 87; 89; 125; 126; 127; 141; 147; 152
space, vii; ix; x; 2; 6; 22; 23; 37; 40; 44; 46; 49; 52; 53; 62; 65; 66; 71; 72; 73; 81; 82; 83; 84; 85; 88; 90; 91; 92; 93; 94; 95; 96; 97; 98; 99; 100; 101; 102; 103; 104; 105; 106; 107; 108; 109; 115; 116; 126; 134; 139; 142; 150; 163; 164; 165; 166; 171; 173
space-time, 106; 107; 108
spatial density, xi; 173
spatial object, 5; 125
spatial point, 60; 72; 82; 83; 87; 89; 90; 92; 95; 118
spatially between, vii; xi; 22; 111; 120; 121; 122; 123; 124; 125; 128; 129; 130; 131; 132; 133; 134; 173
spatially extended, vii; x; 41; 48; 65; 72; 79; 80; 81; 82; 85; 86; 87; 90; 92; 94; 95; 103; 109; 111; 113; 115; 118; 119; 120; 121; 128; 131; 132; 136; 137; 138; 140; 142; 145; 150; 159; 167; 170; 171
spatio-temporal continuity, 65
species, 29; 30; 68; 74; 76; 113; 140
Specimen Dynamicum, 64
spectrum, 27; 29; 30; 31; 35; 36; 37; 48; 66; 73; 76
Spinoza, B., 1; 176
spirits, 28
sprace, 115
structure, 7; 15; 18; 19; 33; 42; 43; 52; 56; 67; 71; 72; 75; 89; 106; 111; 113; 115; 118; 119; 125; 127; 131; 132; 135; 156; 165; 170; 171; 172
stupor, 28; 84
subset, 52; 72; 125; 138; 162; 173
substance, 6; 7; 8; 9; 10; 11; 12; 13; 25; 27; 28; 30; 32; 34; 44; 46; 47; 48; 58; 59; 62; 78; 79; 80; 81; 82; 83; 86; 87; 88; 89; 90; 95; 97; 99; 102; 108; 127; 146; 147; 157; 159; 160; 161; 162; 163; 166; 168; 170
substantial atoms, 60; 61; 62
substantial units, ix; 20; 45; 61; 66; 79; 81; 83; 85; 87; 88; 90; 91; 92; 94; 95; 102; 131; 132; 162; 168
substantivalism, 108
substratum, 107; 108; 132
succession, 99; 134; 138
Supreme Being, 31
Swammerdam, J., 77

T

Theodicy, 158; 159; 176
theology, 159
Theseus, 33
Thomasius, J., 90
thought, 1; 2; 3; 5; 11; 18; 19; 21; 24; 25; 27; 28; 30; 31; 32; 36; 37; 45; 51; 61; 62; 64; 77; 78; 82; 83; 85; 88; 90; 91; 94; 95; 97; 98; 101; 103; 110; 116; 117; 118; 119; 124; 134; 135; 137; 142; 143; 145; 149; 151; 153; 156; 158; 162; 163; 164; 166; 170; 171; 172
time, vii; ix; x; 1; 2; 6; 12; 24; 27; 29; 31; 32; 34; 35; 37; 44; 45; 46; 49; 51; 53; 57; 60; 62; 65; 66; 73; 79; 81; 83; 86; 88; 90; 93; 94; 95; 96; 97; 99; 100; 102; 104; 105; 106; 107; 108; 116; 117; 118; 119; 126; 131; 133; 134; 135; 136; 138; 139; 140; 142; 143; 146; 150; 152; 156; 161; 162; 163; 164; 165; 166; 170; 173
topological space, 52
transcendental number, 71; 72
transmigration, 67; 68
true infinite, 159; 162

U

unconscious perceptions, ix; 14; 23; 30; 31; 32; 38; 41; 68
understanding, ix; 10; 11; 12; 97
uninterruptedness, x; 102; 103; 116; 117; 136; 137; 149; 150; 163; 166; 168; 169; 170; 171; 172
universe, 8; 26; 40; 45; 46; 60; 74; 89; 95; 96; 101; 106; 126; 127; 136; 137; 138; 146; 147; 165; 167

V

vacuum, 7; 11; 22; 40; 59; 62; 65; 83; 84; 114; 115; 139
variable, x; 51; 52; 53; 128; 133; 153; 157
Varignon, P., 36; 74; 77
visual sense perception, 38; 42
void, 88
Volder, B. De, 7; 25; 39; 50; 53; 54; 97; 103; 116; 117; 149; 150; 163; 180

W

well-founded phenomenon, 33; 81; 93; 132; 142
well-foundedness, 6; 112; 114; 142; 143; 153
Wagner, R. C., 27; 33
Whitehead, A. N., 144; 180; 181
Whiteheadian Model, xi; 144
windowless, 11; 25; 121; 126
Wolff, C., 106; 178
world, ix; 1; 2; 3; 4; 6; 7; 8; 9; 10; 11; 12; 14; 17; 18; 20; 23; 24; 25; 27; 31; 33; 35; 37; 39; 40; 42; 43; 44; 45; 46; 50; 53; 54; 55; 56; 57; 58; 59; 60; 61; 62; 64; 65; 67; 68; 69; 79; 82; 84; 85; 86; 87; 90; 95; 96; 97; 100; 101; 106; 108; 109; 110; 113; 115; 118; 119; 120; 121; 122; 123; 126; 127; 130; 131; 132; 133; 136; 137; 138; 139; 140; 142; 143; 144; 145; 146; 147; 148; 150; 151; 152; 153; 154; 155; 156; 157; 159; 167; 168

MIX
Papier aus verantwortungsvollen Quellen
Paper from responsible sources
FSC® C105338

FSC
www.fsc.org

If you have any concerns about our products,
you can contact us on
ProductSafety@springernature.com

In case Publisher is established outside the EU,
the EU authorized representative is:
Springer Nature Customer Service Center GmbH
Europaplatz 3, 69115 Heidelberg, Germany

Printed by Libri Plureos GmbH
in Hamburg, Germany